Managing Quality

For Mary, Glynis and Heather

Managing Quality

Desmond Bell, Philip McBride and George Wilson

Published in association with the Institute of Management

LONDON AND NEW YORK

First published 1994 by Butterworth-Heinemann

2 Park Square, Milton Park, Abingdon, Oxon OX14 4RN
711 Third Avenue, New York, NY 10017, USA

Routledge is an imprint of the Taylor & Francis Group, an informa business

First issued in hardback 2016

British Library Cataloguing in Publication Data
Bell, Desmond
Managing Quality – (Institute of Management Diploma Series)
I. Title II. Series

ISBN 978-0-7506-1823-6 (pbk)
ISBN 978-1-138-15594-7 (hbk)

Typeset by Keyword, Wallington, Surrey

Contents

About the authors

Desmond A. Bell is Company Total Quality Manager at Short Brothers plc. He joined Shorts in 1979 as a technical apprentice and on completion of his training worked as a design engineer. In 1988 he joined Shorts' Total Quality Centre where he contributed to the development of the company's total quality strategy, developed and delivered total quality training programmes for all employees and provided support to the senior management team in steering the total quality initiative. In 1992 he was appointed Shorts Senior Fellow in Quality Management in the Shorts Quality Unit at the Centre for Executive Development of the Ulster Business School. His role was to promote and develop quality and total quality management in the local business community. He is a member of the Royal Aeronautical Society, holds a degree in engineering and a masters degree in engineering business management. He has been trained as an assessor for both BS5750 and the European Quality Award and is a licensed EFQM trainer.

Philip S. McBride is Director; Quality, Strategy and Business Development at Shorts Brothers plc. Having graduated in pure and applied physics from Queens University, Belfast, he joined Short Brothers plc in 1985 as a research engineer conducting research into advanced methods and technologies for use in the future products of the company's Defence Systems Division. In 1989 he was seconded from Research to the Total Quality Centre to help develop and deliver the total quality initiative across the company. In 1990, as a means of furthering links with academic institutions, he was seconded to the University of Ulster as Shorts Senior Fellow in Quality Management with the task of introducing total quality onto various degree courses, conducting research and providing a range of quality courses for business and commerce in Northern Ireland. In August of 1992 he returned to Shorts to take up his current post.

George Wilson is the Management Development Advisor at the Ulster Business School. Before joining the Business School, he was responsible for implementing ISO 9000 in a manufacturing company. At present he teaches on training and development and total quality issues at undergraduate and postgraduate levels. His role also involves providing training for companies in the areas of ISO implementation, leadership and teambuilding, benchmarking, and developing quality programmes in the public sector. He has also been instrumental in developing an integrated TQM suite of programmes which aims to enhance the quality expertise of the local business community. He has been trained as an assessor for the European Quality Award and is a licensed EFQM trainer.

Series adviser's preface

This book is one of a series designed for people wanting to develop their capabilities as managers. You might think that there isn't anything very new in that. In one way you would be right. The fact that very many people want to learn to become better managers is not new, and for many years a wide range of approaches to such learning and development has been available. These have included courses leading to formal qualifications, organizationally-based management development programmes and a whole variety of self-study materials. A copious literature, extending from academic textbooks to sometimes idiosyncratic prescriptions from successful managers and consultants, has existed to aid – or perhaps confuse – the potential seeker after managerial truth and enlightenment.

So what is new about this series? In fact, a great deal – marking in some ways a revolution in our thinking both about the art of managing and also the process of developing managers.

Where did it all begin? Like most revolutions, although there may be a single, identifiable act that precipitated the uprising, the roots of discontent are many and long established. The debate about the performance of British managers, the way managers are educated and trained, and the extent to which shortcomings in both these areas have contributed to our economic decline, has been running for several decades.

Until recently, this debate had been marked by periods of frenetic activity – stimulated by some report or enquiry and perhaps ending in some new initiatives or policy changes – followed by relatively long periods of comparative calm. But the underlying causes for concern persisted. Basically, the majority of managers in the UK appeared to have little or no training for their role, certainly far less than their counterparts in our major competitor nations. And there was concern about the nature, style and appropriateness of the management education and training that was available.

The catalyst for this latest revolution came in late 1986 and early 1987, when three major reports reopened the whole issue. The 1987 reports were *The Making of British Managers* by John Constable and Roger McCormick, carried out for the British Institute of Management and the CBI, and *The Making of Managers* by Charles Handy, carried out for the (then) Manpower Services Commission, National Economic Development office and British Institute of

Management. The 1986 report, which often receives less recognition than it deserves as a key contribution to the recent changes, was *Management Training: context and process* by Iain Mangham and Mick Silver, carried out for the Economic and Social Research Council and the Department of Trade and Industry.

It is not the place to review in detail what the reports said. Indeed they and their consequences are discussed in several places in this series of books. But essentially they confirmed that:

- British managers were undertrained by comparison with their counterparts internationally.
- The majority of employers invested far too little in training and developing their managers.
- Many employers found it difficult to specify with any degree of detail just what it was that they required successful managers to be able to do.

The Constable/McCormick and Handy reports advanced various recommendations for addressing these problems, involving an expansion of management education and development, a reformed structure of qualifications and a commitment from employers to a code of practice for management development. While this analysis was not new, and had echoes of much that had been said in earlier debates, this time a few leading individuals determined that the response should be both radical and permanent. The response was coordinated by the newly-established Council for Management Education and Development (now the National Forum for Management Education and Development (NFMED)) under the energetic and visionary leadership of Bob (now Sir Bob) Reid of Shell UK (now chairman of the British Railways Board).

Under the umbrella of NFMED a series of employer-led working parties tackled the problem of defining what it was that managers should be able to do, and how this differed for people at different levels in their organizations; how this satisfactory ability to perform might be verified; and how an appropriate structure of management qualifications could be put in place. This work drew upon the methods used to specify vocational standards in industry and commerce, and led to the development and introduction of competence-based management standards and qualifications. In this context, competence is defined as the ability to perform the activities within an occupation or function to the standards expected in employment.

It is this competence-based approach that is new in our thinking about the manager's capabilities. It is also what is new about this series of books, in that they are designed to support both this new structure of management standards, and of development activities based on it. The series was originally commissioned to support the Institute of

Management's Certificate and Diploma qualifications, which were one of the first to be based on the new standards. However, these books are equally appropriate to any university, college or indeed company course leading to a certificate in management or diploma in management studies.

The standards were specified through an extensive process of consultation with a large number of managers in organizations of many different types and sizes. They are therefore employment based and employer supported. And they fill the gap that Mangham and Silver identified – now we do have a language to describe what it is employers want their managers to be able to do – at least in part.

If you are engaged in any form of management development leading to a certificate or diploma qualification conforming to the national management standards, then you are probably already familiar with most of the key ideas on which the standards are based. To achieve their key purpose, which is defined as achieving the organization's objectives and continuously improving its performance, managers need to perform four key roles: managing operations, managing finance, managing people and managing information. Each of these key roles has a sub-structure of units and elements, each with associated performance and assessment criteria.

The reason for the qualification 'in part' is that organizations are different, and jobs within them are different. Thus the generic management standards probably do not cover all the management competences that you may need to possess in your job. There are almost certainly additional things, specific to your own situation in your own organization, that you need to be able to do. The standards are necessary, but almost certainly not sufficient. Only you, in discussion with your boss, will be able to decide what other capabilities you need to possess. But the standards are a place to start, a basis on which to build. Once you have demonstrated your proficiency against the standards, it will stand you in good stead as you progress through your organization, or change jobs.

So how do the new standards change the process by which you develop yourself as a manager? They change the process of development, or of gaining a management qualification, quite a lot. It is no longer a question of acquiring information and facts, perhaps by being 'taught' in some classroom environment, and then being tested to see what you can recall. It involves demonstrating, in a quite specific way, that you can do certain things to a particular standard of performance. And because of this, it puts a much greater onus on you to manage your own development, to decide how you can demonstrate any particular competence, what evidence you need to present, and how you can collect it. Of course, there will always be people to advise and guide you in this, if you need help.

But there is another dimension, and it is to this that this series of books is addressed. While the standards stress ability to perform, they do not ignore the traditional knowledge base that has been associated with 'management studies'. Rather, they set this in a different context. The standards are supported by 'underpinning knowledge and understanding' which has three components:

- Purpose and context, which is knowledge and understanding of the manager's objectives, and of the relevant organizational and environmental influences, opportunities and values.
- Principles and methods, which is knowledge and understanding of the theories, models, principles, methods and techniques that provide the basis of competent managerial performance.
- Data, which is knowledge and understanding of specific facts likely to be important to meeting the standards.

Possession of the relevant knowledge and understanding underpinning the standards is needed to support competent managerial performance as specified in the standards. It also has an important role in supporting the transferability of management capabilities. It helps to ensure that you have done more than learned 'the way we do things around here' in your own organization. It indicates a recognition of the wider things which underpin competence, and that you will be able to change jobs or organizations and still be able to perform effectively.

These books cover the knowledge and understanding underpinning the management standards, most specifically in the category of principles and methods. But their coverage is not limited to the minimum required by the standards, and extends in both depth and breadth in many areas. The authors have tried to approach these underlying principles and methods in a practical way. They use many short cases and examples which we hope will demonstrate how, in practice, the principles and methods, and knowledge of purpose and context plus data, support the ability to perform as required by the management standards. In particular we hope that this type of presentation will enable you to identify and learn from similar examples in your own managerial work.

You will already have noticed that one consequence of this new focus on the standards is that the traditional 'functional' packages of knowledge and theory do not appear. The standard textbook titles such as 'quantitative methods', 'production management', 'organizational behaviour' etc. disappear. Instead, principles and methods have been collected together in clusters that more closely match the key roles within the standards. You will also find a small degree of overlap in some of the volumes, because some principles and methods support several of the individual units within the standards. We hope you will find this useful reinforcement.

Having described the positive aspects of standards-based management development, it would be wrong to finish without a few cautionary remarks. The developments described above may seem simple, logical and uncontroversial. It did not always seem that way in the years of work which led up to the introduction of the standards. To revert to the revolution analogy, the process has been marked by ideological conflict and battles over sovereignty and territory. It has sometimes been unclear which side various parties are on – and indeed how many sides there are! The revolution, if well advanced, is not at an end. Guerrilla warfare continues in parts of the territory.

Perhaps the best way of describing this is to say that, while competence-based standards are widely recognized as at least a major part of the answer to improving managerial performance, they are not the whole answer. There is still some debate about the way competences are defined, and whether those in the standards are the most appropriate on which to base assessment of managerial performance. There are other models of management competences than those in the standards.

There is also a danger in separating management performance into a set of discrete components. The whole is, and needs to be, more than the sum of the parts. Just like bowling an off-break in cricket, practising a golf swing or forehand drive in tennis, you have to combine all the separate movements into a smooth, flowing action. How you combine the competences, and build on them, will mark your own individual style as a manager.

We should also be careful not to see the standards as set in stone. They determine what today's managers need to be able to do. As the arena in which managers operate changes, then so will the standards. The lesson for all of us as managers is that we need to go on learning and developing, acquiring new skills or refining existing ones. Obtaining your certificate or diploma is like passing a mile post, not crossing the finishing line.

All the changes and developments of recent years have brought management qualifications, and the processes by which they are gained, much closer to your job as a manager. We hope these books support this process by providing bridges between your own experience and the underlying principles and methods which will help you to demonstrate your competence. Already, there is a lot of evidence that managers enjoy the challenge of demonstrating competence, and find immediate benefits in their jobs from the programmes based on these new-style qualifications. We hope you do too. Good luck in your career development.

Paul Jervis

Preface

The failure to fully understand and successfully introduce total quality is a problem that has existed for some time. The absence of a shared vision of total quality has resulted in many organizations failing to achieve the full benefit that could be accrued if total quality was to be adopted as the fundamental approach to managing the business.

A total quality organization is one that moves beyond short-term goals when identifying its capabilities. Such an organization has processes or methodologies behind its results – processes that are continually reviewed for their effectiveness and appropriateness to changing situations.

A commitment to continuous improvement also exists and future performance is benchmarked against competitors rather than last year's achievements. There is a realization that this level of performance requires a high rate of change and that this can only be achieved through using and developing the capabilities of all employees. The objective of this text is to assess the knowledge and understanding requirements of the manager involved in the development of a total quality programme. The basic themes of the text are:

- How to maintain and improve service and product operations.
- How to implement and monitor a quality assurance system.
- How to ensure key and support processes are identified, reviewed and revised to ensure continuous improvement of the organization's activities.
- How to develop teams and individuals.
- How to plan, allocate and evaluate work carried out by teams.
- The planning, control and optimization of financial resources.
- Obtaining, evaluating and organizing information.
- The necessity to exchange information to solve problems and make decisions.
- How to conduct an appraisal of progress towards total quality.

In order to demonstrate to the manager that the management of quality is a strategic issue which permeates all aspects of organizational performance the individual chapters will develop the basic themes.

Chapter 2, which focuses on quality management systems, reviews how standards of operation are established, how and by whom standards are monitored, the role of the ISO 9000 certification in process management and the process for audit.

Having established the necessity to work to a quality system Chapter 3 reviews how to determine capability for quality by using feedback from customers and suppliers in:

1 determining standards of operation and targets for improvement and
2 identifying, managing and reviewing key processes.

Once capability is determined, Chapter 4 considers how the information gathered can be used to assess how principles of design and operating philosophies are discovered and utilized.

Chapter 5 considers how the need to continuously review process capability in order to satisfy customer requirements can require the use of statistical measurement. Examples of statistical measurement illustrate how current targets for improvement are related to past achievement.

The management of financial resources is another example of measurement. Chapter 6 considers the definition and use of quality cost concepts and the establishment of a quality cost system.

As indicated the total quality approach to management necessitates development beyond the functional perspective. Traditionally, quality was considered relevant for only the management of product-related operations. Chapters 7–9 illustrate that the management of people is an important indicator of effective management. Chapter 7 considers how the organization will structure itself to facilitate quality improvement and how managers need to review and improve human resources management. The objective of Chapter 8 is to illustrate that, as continuous improvement is dependent on teamwork, there is a need to promote involvement and develop understanding of what constitutes effective teamwork. The success of teams is determined by effective team members having the necessary quality improvement 'toolkit' to facilitate improvement activity. The importance of establishing and implementing training plans and reviewing their effectiveness is considered in Chapter 9.

The concluding chapter, Chapter 10, recognizes that the transformation of quality into a strategic business planning dimension is dependent on ongoing measurement and appraisal. In order to enable organizations to assess progress towards total quality the EFQM self-assessment model is discussed.

Desmond Bell, Philip McBride and George Wilson

Acknowledgements

The authors are grateful to the many people and organizations who have contributed to the case studies in this book. In particular we would like to acknowledge the time and effort committed by: Paula Kane, Raymond Kane; Robert Beckett, Valpar Industrial Ltd; Short Brothers plc.

We also wish to acknowledge the invaluable help and support of Hazel Cameron, Ann McFarlane and Lynda Malcomson in the preparation and completion of this text.

1 Quality as a strategic issue

Introduction

In the period between 850 and 750 BC a philosopher and the author of the biblical text Ecclesiastes suggested that 'there is no new thing under the sun'. The implication is that on Earth, anything that will happen in the future has already occurred, to a certain extent, in the past. Wars will continue, as they have done since recorded history began, the only difference being their scale and the technology involved, diseases will continue to be cured and unfortunately established and again the difference will probably be in the scale of those affected. To the same extent business acumen will continue to separate the best from the also-rans. Whether in 800 BC as a trader in high-quality spices or today as a multinational organization involved in many spheres of industry and commerce, quality is and always has been the distinguishing factor.

Today, over 2500 years later the 'no new thing under the sun' statement still holds true, this time with regard to quality management. It is said that focusing on our customers is simply putting into practice the marketing principles we learned about 20 years ago. Some argue that process control is what we learned from Shewart in the 1920s, others say that doing things right first time is something all organizations strive to do in any case.

Even if these statements are individually true, and the component parts are not new, the scale, integration and application of these component parts and more is most certainly revolutionary in business terms. It is this scrutiny applied to everything we do,

> from the way we design our products to the way we support our customers through after-sales service, from the way we manage our people to the way we manage our finances, from the way we develop our organizational strategies to the way we handle information,

that describes the management philosophy which has become known as total quality management.

The development of quality

Traditionally, quality has been associated with the product that a customer receives. This is very laudable but placing a high-quality product or service in the marketplace does not, on its own, guarante either sales or business success. To this end, quality and total qualit must be placed in the context of a business environment.

One of the main difficulties evident in the field of quality manage ment is the variety of terms employed. Many are ill-defined and ar used both interchangeably and inconsistently. This confusion can b reduced by considering the following generic terms: quality contro quality assurance and total quality management.

Quality control

Quality control (QC) concerns the techniques and activities whic sustain quality to specified requirements. This system has for man decades been the traditional understanding of the term quality.

It refers to the practical means of securing product or service qualit as laid out in a product specification. Quality control may be viewed a a subset of quality assurance, although, chronologically, quality contro was used first. The basis of quality control is inspection. An importan outcome of this statement is that quality control is an 'after the fact activity which measures product that has not been produced to cus tomer satisfaction. In other words, defects are detected through post production inspection by a QC system and not prevented. Richarc Schonberger[1] refers to this as the 'death certificate approach'.

Typical activities in a QC environment include:

- Determination of inspection points
- Inspection method development
- Data collection and analysis
- Prevention of chronic problems

Since QC is not essentially a prevention-based system, the process o measuring, examining, testing and comparing with relevant standard must be accurate and precise. Unfortunately, while many organization recognize quality as a major factor affecting customer choice they alsc believe that traditional quality control techniques will improve quality standards.

However, employing more inspectors and detecting more defect does not promote or improve quality in the medium to long term The inspection process is costly and adds no value to the organizatior or the product. Even if all defective parts are removed before delivery to the customer the cost of incurring the defective products in the firs place will be passed on to the customer in some way.

Quality assurance

Quality assurance (QA) recognizes that inspection is not enough in itsel to remedy quality problems. It focuses on procedure compliance and product conformity to specification through product and operations management tracking. Today, QA has become synonymous with the British Standard BS 5750 or its international equivalent ISO 9000. BS 5750 defines a quality systems standard, and this is important in that it relates to a *system* and not a *product*. The standard sets out a framework by which a management system can be implemented such that the needs of the customers are fully met.

Quality assurance is based on the principle of prevention of quality problems rather than detection of these problems as it is in quality control. Inspection and quality control are still important tools, but we need more planned and systematic actions than these in order to prevent quality problems recurring. Quality assurance activities consider:

- How an organization develops policy in respect of quality
- The allocation of responsibilities within the organizational structure
- Procedures used to carry out the needs of the business
- The standards to be attained in the workplace
- The documentation required to demonstrate both the operation and maintenance of the system and the attainment of quality.

Total quality management

As already implied, the word 'quality' has traditionally been associated with the final product or service which is subsequently offered to a customer or client. Hence, definitions such as that of Joseph Juran,[2] who describes quality as being 'fitness for use', or that of Philip Crosby,[3] who defines quality as 'conformance to specification'.

A modest expansion on the concept of quality suggests that the product or service should not only be fit for use but should also be delivered to the customer when he or she has asked for it and also at a cost which the customer is willing to bear. While the above definitions obviously relate to the final product or service, the underlying logic suggests that for a product or service to be delivered on time, at cost and at the customer-desired level of quality, all the processes and support processes which in effect produce the product or service must be both efficient and effective. This expansion of the concept of quality thus includes aspects internal to the business rather than simply the traditional external attributes of the organization. Consequently, if total quality management is about satisfying customer requirements then the implementation of total quality must take into account the idea of the internal customer.

If we think of each person in an organization as having three distinct roles it becomes easier to understand the internal customer concept. First, an individual is an operator since he or she performs some value-added activity on materials or information. Second, the individual is a supplier as he or she passes the processed information or materials on to someone else in the value chain for further processing. Third, the individual is a customer as he or she receives information or materials on which to perform value-added activities. Therefore within any organization there are a series of customer–supplier interfaces which need to be managed.

So far we have established that the external customer demands products and services on time, at cost, at the appropriate quality and which add value to his or her operation. These external customer requirements are satisfied by the way in which we manage our internal customer-supplier relationships. Moreover, if we think of the organization as an entity which transforms materials and information into products and services for the external customer, then we must also include the external supplier as an important element in the value and quality chain. This being the case, we can now begin to understand some of the contextual differences between quality in its traditional sense and total quality as it is recognized today.

The traditional understanding of quality, whether it be quality control or quality assurance, tends to focus directly on the product or service provided. However, total quality focuses on the interactions of the external customer, the external supplier, stockholders, society at large and the organization itself and specifically on the effective and efficient management of the processes which satisfies the needs of this extended enterprise.

European framework for total quality management

During the 1980s many organizations in the West began to realize that their survival depended on the degree to which they achieved quality in products and services and indeed within all areas of the extended enterprise. Today, quality has become the competitive edge in many areas of industry and commerce. Realizing that quality management will become increasingly important, fourteen leading Western European businesses formed the European Foundation for Quality Management (EFQM) in 1988. The key roles of this organization are to:

- Accelerate the acceptance of quality as a strategy for global competitive advantage
- Stimulate and assist the deployment of quality-improvement activities.

The European Quality Award was established by the EFQM in 1991, in conjunction with the European Organization for Quality and the European Commission. The European Quality Model itself provides a meaningful mechanism for self-appraisal. This appraisal involves the regular and systematic review of the organization's activities and results.[4]

By way of making the framework more easily understandable and applicable to almost any business the model describes nine key areas, or criteria, which are important to all organizations (Figure 1.1). The first five criteria are:

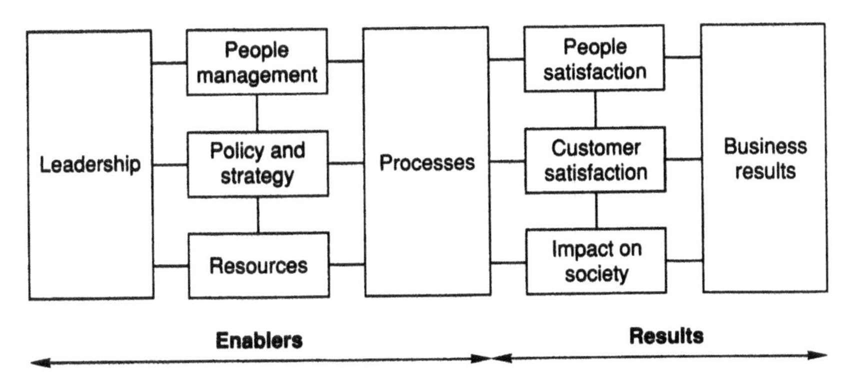

Figure 1.1 *The European TQM model for self-appraisal*

- Leadership
- People management
- Policy and strategy
- Resources
- Processes

These criteria are termed Enablers and are concerned with *how* results are achieved within an organization. The final four criteria:

- People satisfaction
- Customer satisfaction
- Impact on society
- Business results

are called Results and are concerned with *what* the organization has achieved and is achieving.

The framework indicates that customer satisfaction, people satisfaction and impact on society are achieved through leadership driving policy and strategy, people management, resources and processes, leading ultimately to excellence in business results. Thus, if the words 'total quality' do not fit into a particular organization they may be substituted for 'business excellence', as this model simply proposes a framework for managing a business in an excellent way in terms of today's thinking.

The self-appraisal guidelines show how each criterion may be split into a series of criterion parts covering various aspects of the parent criteria. Each subsequent criterion part has a series of areas to address, which may or may not be considered by an organization.

Total quality is an all-embracing management philosophy and should not be viewed as an add-on activity. It cannot be seen as an additional task for the manager. Rather, it must form an integral part of the manager's role and indeed part of all others in the organization. For this to be the case total quality must form part of the fabric of the organization. Total quality is not what we do. It is what we are.

Consequently, quality management must be used as a competitive weapon within the organizational strategy and the organizational strategy must be developed in a total quality way. Only then will an organization be seen to be striving for world-class status.

These criteria and associated sub-criteria will be considered in more detail in Chapter 10. However, a brief review of the elements will serve to illustrate how the integrative nature of the model requires a new perspective on how to manage quality.

Quality strategy and policy

Leadership

The behaviour of all managers is driving the organization towards total quality. A total quality approach should demonstrate:

- Visible involvement in leading total quality
- Recognition and appreciation of the efforts and successes of individuals and teams
- Support of total quality by provision of appropriate resources and assistance
- Involvement with customers and suppliers.

Policy and strategy

The organization's mission, values, vision and strategic direction and the ways in which the organization achieves them:

- How policy and strategy are based on the concept of total quality
- How policy and strategy are formed on the basis of information that is relevant to total quality
- How policy and strategy are the basis of business plans
- How policy and strategy are communicated
- How policy and strategy are regularly reviewed and improved.

People management

The management of the organization's people. How the skills and capabilities of the people are preserved and developed through recruitment, training and career progression:

- How people and teams agree targets and continuously review performance
- How the involvement of everyone in continuous improvement is promoted and people are empowered to take appropriate action.

esources

The management, utilization and preservation of resources:

- Financial resources
- Information resources
- Material resources
- Technology resources.

rocesses

The management of all the value-adding activities within the organization:

- How processes critical to the success of the organization are identified
- How the organization systematically manages its processes
- How processes' performance measurements, along with all relevant feedback, are used to review processes and to set targets for improvement
- How the organization stimulates innovation and creativity in process improvement
- How the organization implements process change and evaluates the benefits.

Customer satisfaction

- What the perception of external customers is of the organization and of its products and services.

People satisfaction

- What the employee's feelings are about their organization.

Impact on society

- What the perception of the organization is among society at large. This includes views of the organization's approach to quality of life, the environment and the preservation of global resources.

Business results

- What the organization is achieving in relation to its planned business performance (financial and non-financial measures).

The objective of this text is to assess the knowledge and understanding requirements of the manager involved in the development of a total quality programme. The basic themes of the text are:

- How to maintain and improve service and product operations
- The implementation of changes in services, products and systems
- How to ensure that key and support processes are identified, reviewed and revised to ensure continuous improvement of the organization's activities
- How to develop teams and individuals
- How to plan, allocate and evaluate work carried out by teams
- How to create and maintain affective working relationships
- The planning, controlling and optimization of financial resources
- Obtaining, evaluation of and organizing of information
- The necessity to exchange information to solve problems and make decisions.

The integration of a total quality initiative within the overall business planning process is dependent on the formulation of:

- An adequate mission
- Adequate policies
- Adequate associated objectives

The adequacy of an organization's mission, policies or objectives is dependent on the relevance of information that is gathered and employed in the formulation of strategy and policy. It is essential to develop mechanisms which will clearly identify the needs and expectations of existing/potential customers. The use of surveys, questionnaires and personal visits ensure that relevant information is gathered. Consideration should also be given to involving customers as regards new product development and feedback on possible strategic development.

An assessment of customer requirements and current market trends enables the identification of:

1 Required inputs to satisfy customer requirements
2 Areas to be addressed where a deficiency is identified.

The information gathered enables a strategy to be developed which is based on actual customer input and which outlines the objectives of the quality initiative on a medium- to long-term basis (i.e. 3–5 years). For example, the review of an organization's activities may indicate the following primary elements which form a TQM strategy:

- People
- Technology
- Methodology

Each of the three primary elements can then be examined to identify tools, techniques and concepts which are applicable to the organization. (see Figure 1.2).

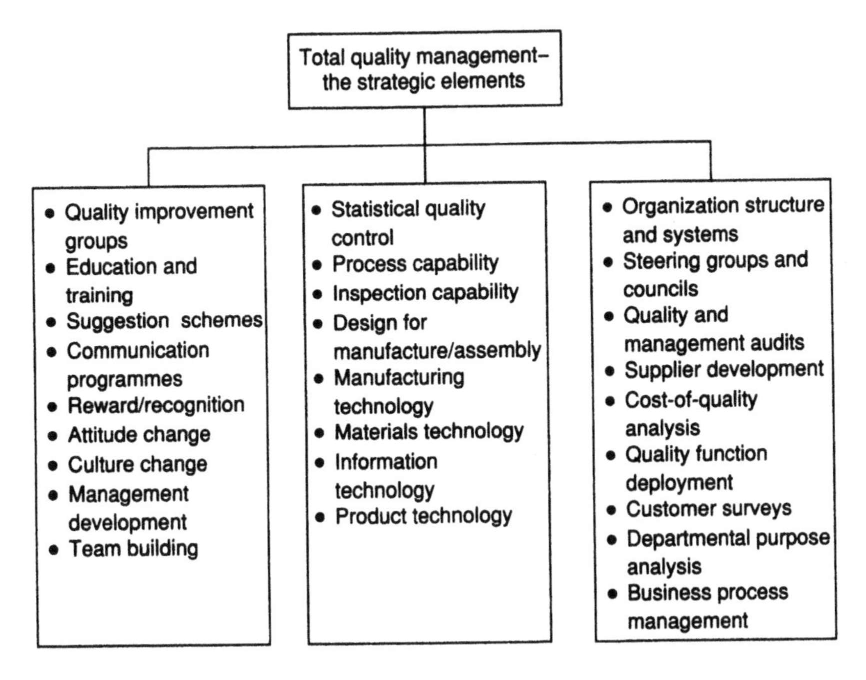

Figure 1.2 *The elements of a total quality management strategy*

A total quality strategy can embrace many objectives:

1 To improve service to the customer
2 To improve business reliability and operating efficiency
3 To develop people-involvement mechanisms
4 To improve company–employee communications
5 To establish clear departmental goals
6 To facilitate an open style of management and team building
7 To achieve accreditation to ISO 9000.

Examples of quality policies

Total quality policy

We believe that our total commitment to continuous improvement will guarantee the future of this company by fulfilling the needs and expectations of our customers and employees in a responsible, professional and more profitable way. Our objectives are:

1 To introduce and maintain a company-wide quality improvement programme
2 To achieve the total commitment of all employees
3 To achieve accreditation to BS 5750
4 To meet the needs and expectations of our internal and external customers
5 To improve communication between our company and its customers.

Customer relationships policy

We believe that we will enjoy successful relationships with our customers by continuously meeting their needs and expectations through our policy of quality and reliability. Our objectives are:

1 To target those customers with whom we wish to develop a mutually profitable business relationship
2 Continuously to develop and promote our reputation as a quality company
3 To develop honest and trusting relationships through interactive communications
4 To develop a customer-satisfaction measurement mechanism which will help to ensure that customers' needs and expectations are met

Employee relationships policy

We believe in providing secure and satisfying employment to all employees in an environment where ability and commitment are recognized and rewarded. Our objectives are:

1 To convince each employee that their whole-hearted participation in the company-wide quality improvement programme is vital to its continuing success
2 To develop and maintain effective and open communication with all employees
3 To identify training needs
4 To provide opportunities for the development of all employees through specific training
5 To create an environment in which ability, commitment and quality of performance are recognized and rewarded
6 To provide a safe working environment
7 To ensure that company procedures and safe working practices are understood and adhered to
8 To continue to provide all employees with the resources necessary to ensure the success of the company.

Monitoring performance

We believe in monitoring the performance of the company and its employees at all levels as part of the process of continuous improvement. Our objectives are:

1 To ensure that all employees understand the company's policies and objectives
2 To establish indicators against which company performance can be measured
3 To agree performance indicators for all employees
4 To establish a system for monitoring the performance of the company and employees
5 To monitor individual performance and abilities
6 To take corrective action.

Business planning policy

We believe that we can profitably meet the needs and expectations of all our customers by providing a quality product and service through a commitment to continuous planning and improvement. Our objectives are:

1 To define the needs and expectations of existing and potential customers by continuous market research
2 To develop a business plan to include marketing, operational and financial strategies
3 To make full use of the resources within the company
4 To promote and enhance the company's image in our chosen market
5 To have and maintain effective leadership within the company.

Each detailed objective can then be developed to incorporate a number of key aspects that should be concentrated on by management and used to guide the tactical implementation in individual departments and functions. Subsequently, annual operating plans can be developed which detail the specific objectives and priorities.

Any objectives identified should focus on meeting the needs of the business as identified in the strategic plans. All total quality activity should be focused on meeting these objectives. In addition to continually gathering information from customers as regards the adequacy of policies and strategies in helping meet their needs, it is also beneficial to receive feedback from employers in terms of how strategies impact upon them. Team meetings and presentations are possible methods of receiving feedback. Indeed, existing methods of communication with employees can be used to formulate, communicate and review strategy. This can be employed as a mechanism for developing the

commitment of employees to the implementation and realization o
strategies.

Summary points

- All quality activity should be focused on meeting the needs of the business.
- Quality control refers to the practical means of securing product or service quality as laid out in a specification.
- Quality assurance is based upon the principle of prevention, rather than detection, of quality problems.
- Total quality is everyone, at all levels, striving to continuously improve the quality of product or service to the customer.
- The European Framework for Total Quality Management provides a meaningful mechanism for the regular and systematic appraisal of an organization's activities and results.

References

1 Schonberger, R. J. (1990), *Building a Chain of Customers*, New York: The Free Press.
2 Juran, Joseph, M. (1979), *Quality Control Handbook*, 3rd edition, New York: McGraw-Hill.
3 Crosby, Philip (1979), *Quality is Free*, New York: McGraw-Hill.
4 European Foundation for Quality Management (1993), *The European Model for Self-Appraisal*.

2 Quality management systems

Introduction

Total quality as a management concept is still evolving. However, the theoretical underpinnings have remained constant and derive from simple definitions that have much wider implications:

- A product is any and all output perceived by a customer
- A process is everything carried out to generate the output
- A customer is the recipient of the output
- A supplier is the provider of the output.

A product is a noun – or a 'what'. It can be a service; a mail order catalogue; a training session; software; the courteousness of a telephonist; a response to a problem. To a customer, perception is reality. Processes are verbs – or 'hows'. They include all the internal operations and sequences of operations used to generate and deliver a product such as designing; word processing; scheduling; planning; writing; writing material. Customers can be either internal or external. Vendors are customers of the purchasing department; individuals are customers of each other in an organizational context, irrespective of status. Suppliers can also be internal or external.

Unlike external customers who do not directly impact on internal processes, problems or costs, suppliers are directly associated with processes. Their primary objective is to produce customer-acceptable products using processes which have a high value-to-cost ratio. The higher the ratio, the more efficient and competitive the supplier will be.

One overriding concept of total quality is the necessity for requirements. A requirement is a parameter that sets limits on either a product or a process. It is usually established by the customer on products and by the supplier for processes. All activities of all enterprises are related to setting or meeting a requirement (Figure 2.1). The objectives of this chapter are to enable the manager to:

- Develop an understanding of how to introduce, develop and evaluate quality assurance systems
- Optimize the use of resources based on criteria derived from customer agreements and requirements
- Identify and encourage appropriate members of staff to assist in the development of quality assurance systems
- Communicate the benefits and results of ensuring quality to enhance employee commitment and customer satisfaction

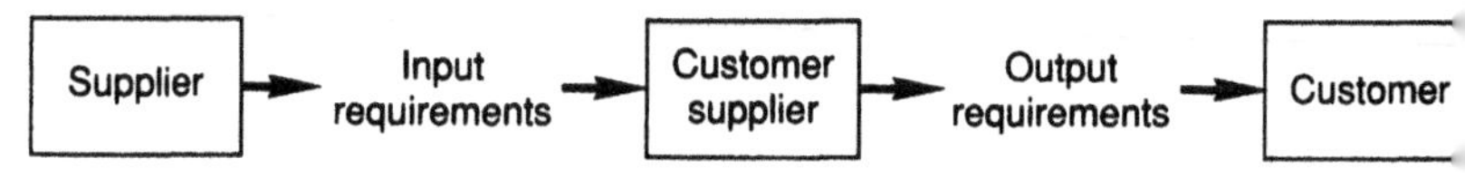

Figure 2.1

- Change services, products and systems in accordance with implementation plans and agreed delivery specifications
- Assess suppliers
- Monitor and record corrective action
- Draw up implementation/modification schedules or plans
- Conduct internal audits.

Quality systems

An organization constitutes an environment where a multitude of tasks are carried out. A process can be defined as the grouping in sequence of all those tasks directed at accomplishing one particular outcome. Therefore every activity becomes part of a process. The interpretation of activities in terms of processes is perhaps the most profound change that occurs during the transformation to quality maturity. The relationships between jobs become apparent – improving processes improves work activities.

Employees are able to develop a unified language and understanding of what their jobs entail. Specific steps in any one particular process can be understood in the context of the overall organizational remit. This improved understanding of activity enables employees to define the capacity of existing processes to satisfy customer requirements. They will be able to focus on errors, waste and other problems, and determine what data will help improve the effectiveness of this collection of tasks (see Chapter 3).

The interpretation of work as processes enables recognition that the quality of output is determined by the quality of input. How well employees can do their jobs depends on the quality of product or services they receive:

- An assembler depends on the quality of product or service received
- Typists depend on those who write the documents
- Everyone depends on the policies, methods, tools and equipment provided by the managers.

If a series of related tasks can be called a process, a group of related processes can then be seen as a *system*. The quality system of an organization is characterized by its network of processes, not only their structure but also responsibilities, procedures and resources that relate to the processes. Ultimate effectiveness of a quality system requires that these processes and their associated responsibilities, procedures and resources be documented and deployed in a consistent manner. An effective quality system, therefore, needs coordination and compatibility of the component processes. This involves strategic planning and setting priorities among the component processes. It is conventional to speak of a quality system as composed of a number of elements, each dealing with a function or group of related functions which need to be carried out by means of processes.

The fundamentals of quality assurance

When a purchaser places an order with a supplier, it is accepted that they know the required standard of the goods or services being ordered. The needs of the purchaser may be:

- Anticipated by the supplier and manufacturer in advance;
- Stated by the purchaser in the form of a specification for the goods; or,
- Stated by the purchaser in the form of a specification of the performance required. In this instance, the provider/manufacturer will draw up the specification for the goods as well as providing the service and/or manufacturing the product.

The needs of the purchaser are referred to as 'specified requirements' and are defined as:

- Requirements prescribed by the purchaser in a contract for material or services
- Requirement prescribed by the supplier that are not subject to direct specification by the purchaser.

Quality assurance is the all-embracing title which covers all functions, from market research through production to field service, which ensures that the customer obtains a product or service which is fit for the purpose.

ISO 9000

Evolution

The evolution of ISO 9000 can be traced back to the 1950s where the initial thrust for quality systems came via the MIL standards in the USA during the construction of the first nuclear submarines and power stations. During the 1960s interest focused on the importance of clearly defined systems – a documented approach to ensure quality. The MIL standards were taken up by NATO members in the procurement of military hardware. A great deal of the basic work carried out in the compiling of standards for quality assurance stems from the NATO standards known as the AQAP series (Allied Quality Assurance Publications). These standards gave detailed requirements, stated the need for the requirements and suggested evaluation questions.

In 1972 the Ministry of Defence Procurement Executive announced a new system of quality assurance which they proposed to adopt for all their purchased materials. The system was based directly on the NATO requirements and the MOD issued a corresponding set of defence standards, the most important being the 05 series (05-21, 05-24 and 05-29). Thus the concept of formal assessment was introduced in the UK by the MOD assessing its main contractors and sub-contractors.

The defence standards were virtually adopted by the British Standards Institution in 1974 with the standard BS 5179 being published.[1] However, it was only a guide, as many sectors of industry had misgivings about simply reissuing a MOD standard as a BSI contractual standard. This position was finally resolved in 1979 with the issue of BS 5750/ISO 9000.[2]

In March 1987 the International Standards for Quality Systems were published (ISO 9000-9004) and in June 1987, BS 5750 was aligned and is now identical to the new ISO. The publication of the ISO 9000 series has brought an international recognition of quality system assessments. The successful adoption by CEN of the EN 29000 series of European Standards, mirroring the ISO 9000 series, also took place in 1987.

What is ISO 9000?

ISO 9000 is a quality system standard that sets out the methods by which a management system, incorporating all the activities associated with quality, can be implemented in an organization to ensure that all the special performance requirements and the needs of the customer are fully met. The standard applies to the quality management systems a company uses, not its product. It is a management system which can be applied to almost every manufacturing or service company.

ISO 9000 describes how you can establish, document and maintain an effective quality system which will demonstrate to your customers that you are committed to quality and are able to supply their quality needs. The standard has been broken down into sections to enable manufacturers to implement it easily and efficiently. It provides a framework for designing your own quality system. In the various clauses the essential requirements of a sound and practical way of working are specified. Once devised, your quality system should be flexible enough to handle any work you undertake. Whereas quality means meeting the needs of your customer, the quality system *makes sure* that you meet the needs of the customer. Quality assurance is a management discipline and a logical progression of activities which influence the quality of the end product.

The series of standards ISO 9001, 9002, and 9003 provide models for three levels of quality assurance.

- ISO 9001 (BS 5750: Part 1) provides the model for an organization which is involved in the management or design as well as in producing its product or service. Thus in service organizations or organizations providing professional services where the service offered is designed to meet the specific needs of the customer, ISO 9000 is the applicable model.
- ISO 9002 (BS 5750: Part 2) is the appropriate model for many manufacturing industries producing standard items or service organizations such as retailing outlets providing a standard service.
- ISO 9003 (BS 5750: Part 3) is only used for organizations whose product is already manufactured and is simply inspected before being supplied (see Figure 2.2).

AQAP 1–3	NATO Requirements for an Industrial Quality Control System
AQAP 4–2	NATO Inspection System Requirements for Industry
AQAP 9–12	NATO Basic Inspection Requirements for Industry
BS 4778: 1986	Part 1 [ISO 8402]: Quality Vocabulary: International Terms
BS 4778: 1979	Part 2 Quality Vocabulary: National Terms
BS 4891: 1972	Guide to Quality Assurance
BS 5750	Quality Systems
1987	Part 1: Specification for Design, Development, Production, Installation and Servicing
1987	Part 2: Specification for Production and Installation
1987	Part 3: Specification for Final Inspection and Test
1990	Part 4: Guide to the Use of BS 5750: Parts 1, 2 and 3
BS 6143	Guide to the Economics of Quality
BS 7229: 1989	Guide to the Quality Systems Auditing
ISO 9001: 1987	Quality Systems – Model for Quality Assurance in Design, Development Production Installation and Servicing
ISO 9002: 1987	Quality Systems – Model for Quality Assurance in Production and Installation
ISO 9003: 1987	Quality Systems – Model for Quality Assurance in Final Inspection and Test
BS 7850: 1992	Total Quality Management

Figure 2.2 *Quality system standards*

How does it operate?

ISO 9000 requires objective evidence at all stages of a process, from raw material purchase to final delivery to the customer, that work is being carried out to agreed standards. In order to achieve that objective, the company has to produce a quality manual which guides its users to relevant documents which indicate what is expected at each stage of production. Often companies elect to draw up a defined quality system in three stages:

1 *Quality manual*: A policy document saying what the company intends to do to ensure that a system exists and will guarantee consistent product quality. This is normally a document which hardly ever changes and which contains little specific detail on how the result is achieved. As such, it can be sent out to customers to give them assurance that they will receive what they request
2 *Departmental procedures*: Detailed procedures which are instructions to operators and inspectors giving instructions on how equipment is to be operated; how to carry out inspections and tests; and how to complete documentation
3 *Specifications*: Specific instructions relating to particular contracts. This would include drawings, specific test procedures, quality plans, etc. These nearly always change from contract to contract.

If a decision is made to implement a quality system it must be borne in mind that the quality system must be long term. If an organization/manufacturer does not have the resources to set up the system itself, a consultancy may be the solution. Most systems exist informally or semi-informally, and it is usually a case of formalizing and documenting the method of operation. If external consultants are employed, what an organization must be wary of is ready-made work processor procedures which may not suit them and their business. Certification bodies do not normally act as consultants, indeed, some government accreditation criteria rule out this activity.

The elements and requirements comprising these three models are (see Figure 2.3):

- Management responsibility
- Quality system
- Contract review
- Design control
- Document control
- Purchasing
- Purchaser supplied product
- Product identification (and traceability) (Parts 1 and 2 only)
- Process control
- Inspection and testing

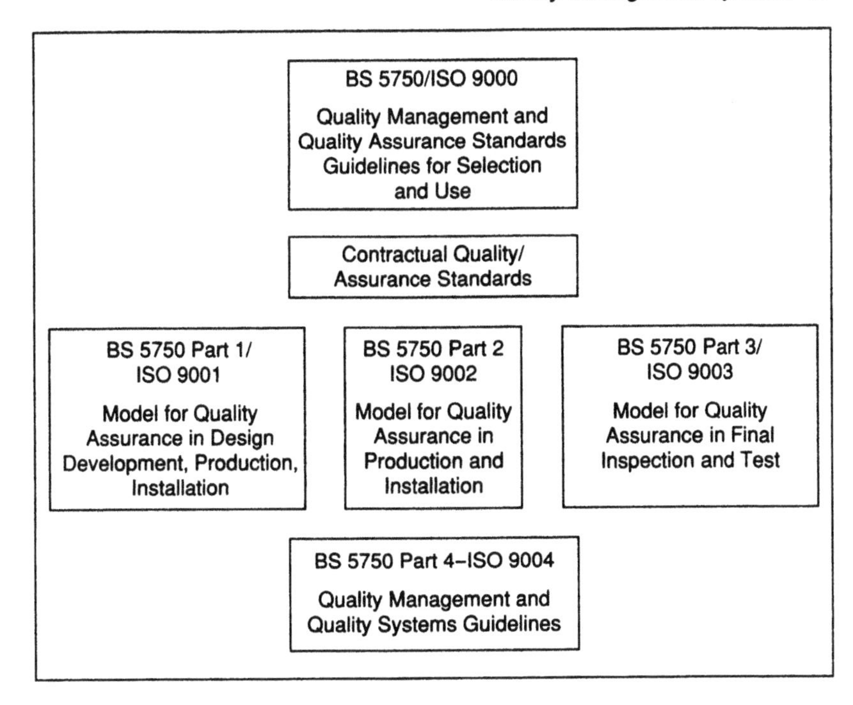

Figure 2.3 *BS 5750/ISO 9000–9004 standards*

- Inspection, measuring and test equipment
- Inspection and test status
- Control of non-conforming product
- Corrective action
- Handling, storage, packing and delivery
- Quality records
- Internal quality audit
- Training
- Servicing
- Statistical techniques

Sub-clause 4.1: Management responsibility

The commitment of an organization to a quality system must be clearly indicated within a declared quality policy which is documented and included as an integral part of the quality policy manual. This normally takes the form of a statement signed by the chief executive or equivalent, which demonstrates commitment to providing products or services than meet the 'fitness for use' definition of quality. Typically, one person is nominated as the management representative for quality. This appointment must have sufficient overall responsibility and authority to coordinate, implement, maintain and monitor the quality system.

Sub-clause 4.2: Quality system

This sub-clause stipulates that an organization must establish a fully documented quality system which, on implementation, ensures that product or service conforms with specific requirements. Typically the quality system comprises:

1 Development of a quality manual
2 Assessment of resources required to attain the desired quality
   - Process facilities
   - Inspection, measuring and test equipment
   - Skill levels
3 Clearly defined acceptance and rejection criteria for all features and requirements of the product or service
4 Quality records to be used for verification purposes.

The quality system should extend beyond traditional quality control activities and include all functional areas.

Sub-clause 4.3: Contract review

The purpose of the contract review is twofold:

1 It provides a record of the customer's requirements, highlighting differences between the customer's statement of requirements and the product or service to be supplied
2 It enables a supplier to assess capability to meet the customer's requirements. Contractual requirements must be evaluated against current resources to ensure that the capability exists to undertake the specific order.

Sub-clause 4.4: Design control

If an organization has responsibility for design and development of a product, a procedure should be documented which details the sequence of design and development activities involved in the manufacture/provision of a new or updated product. It is important that those responsible for design communicate with other functional areas influenced by product design. The input of quality assurance Purchasing, Engineering, Production Planning and Personnel is essential to the coordination of activities which will contribute to the achievement of contractual obligations and delivery schedules.

Suitably qualified personnel should be responsible for carrying out and documenting the following activities:

- Design review meetings
- Testing or demonstrating to prove design
- Applying alternative calculations to the design
- Comparison of new design against existing product.

Sub-clause 4.5: Document control

The documentation associated with the requirements of the quality system must be included in a document control procedure and is composed of:

- Manuals
- Instructions
- Operational documents
- Quality plans
- Specifications and drawings
- Records and reports

Before formal approval and issue, these documents have to be reviewed for adequacy by authorized personnel. Ongoing review is also required to ensure that the current status of documentation is distributed to relevant personnel. Records of distribution must be maintained and review of document adequacy should be included in audit procedures (internal audit – management review).

Sub-clause 4.6: Purchasing

The control of materials and products received by an organization will involve both the purchasing and acceptance procedures to ensure that all specified requirements are being met. Consequently, the evaluation of suppliers and sub-contractors is an essential element in the development of confidence in a supplier. The selection of suppliers and sub-contractors is likely to encompass the following:

- Past performance (e.g. quality, pricing and delivery)
- Adequate facilities
- Supplier's quality system
- Accreditation of supplier to an appropriate quality system.

The company's organization should enable clear communications between the purchasing function and other functional departments such as: design, planning, quality assurance, engineering, maintenance, etc.

Sub-clause 4.7: Purchaser-supplied product

This sub-clause is applicable to those organizations who receive products or materials from a customer from a part of the product ordered by that customer. These customer-supplied goods are, in effect, owned by the customer and procedures should be put in place to ensure that:

- No damage is caused to the product or materials
- The product or materials are clearly identified
- The product or materials are maintained
- The product or materials are stored in a segregated area
- The product or materials are subject to special handling and usage

Sub-clause 4.8: Product identification and traceability

Procedures must be established to enable the identification of the product at all stages in its progress in production, delivery and installation. The documentation associated with identification and traceability include the specification and drawings for a specific contract. During the process of manufacture, works order documentation must accompany each item or batch. Such documents include:

- Quality plans
- Work description sheets
- Parts lists
- Work-in-progress time sheets
- Batch cards

Sub-clause 4.9: Process control

To ensure the effective control of the production process, the process itself must be subject to clear and specific planning. This involves decisions being made on how the contract specification can be translated into the actual product using the resources available. Such resources include:

- Manpower and skill levels
- Process equipment
- Inspection measurement and capability

All process operations should be specified in comprehensive, documented work instructions (see Figure 2.4)

Sub-clause 4.10: Inspection and testing

Inspection and test procedures should be established to provide the highest possible level of confidence at low cost. This should incorporate the cost of passing reject-status work to subsequent process stages. The procedures should, therefore, incorporate the inspection strategy to be adopted and the techniques and equipment to be employed for checking designated product characteristics. As the quality system should be based on prevention, the inspection activities

Operations	Work instructions	Inspection
Identify material Detail activities List tools and equipment Operational monitoring Set-up and maintenance of equipment Judgement criteria for workmanship	Quality plan Procedures Drawings Specifications Standards	Test equipment required Calibration of test/ inspection equipment List inspection methods Criteria for passing or failing inspection tests Sampling techniques

Figure 2.4 *Process operations and instructions*

outlined are related to the control of the process, and not simply to final inspection or rejection of the product.

Parts 1 and 2 of ISO 9000 consider the testing and inspection of materials at three stages:

- Incoming goods and raw materials
- Processing of the products
- Final inspection and testing.

Sub-clause 4.11: Inspection, measurement and test equipment

As product quality conformance is indicated by inspection and test measurements it is essential that test equipment be accurate. Control of inspection and testing requires:

- Adequately trained inspection personnel
- Suitably calibrated and controlled equipment.

The frequency of calibration for each item of equipment is stated in a calibration programme which is an equipment status document highlighting the dates when calibration is due to take place and when it was carried out.

Sub-clause 4.12: Inspection and test status

A procedure is required to ensure that the inspection or test status of items or batches of material or product is clearly identified. There should be control of items revealed by inspection/test to be out of compliance with specified requirements. Authorised stamping, tags, labels or various other methods of identification may be used to denote:

- Not inspected or tested status
- Approved status

- Concessionary approval status (dependent on gravity of non-compliance)
- Rejected status.

Sub-clause 4.13: Control of non-conforming product

Procedures are required to prevent the inadvertent use or installation of products which do not conform to the specification – their status should be identified accordingly. A non-conformance report should communicate the nature of the problem to the management representative for quality and other relevant parties. Categories of non-conformance are:

- Rework or repair to conform to specified requirements
- Regraded
- 'Use as it is' – authorized by relevant personnel/customer concession
- Rejected.

Sub-clause 4.14: Corrective action

When a product is found not to conform to the specified requirements a procedure is required to ensure documentation, implementation and verification of the subsequent corrective action. A causal investigation by authorized personnel using associated documentation determines the appropriate corrective action. When agreement is reached relevant to the corrective action, this is documented formally on the appropriate section of a non-conformance or corrective action report. The remedial action agreed upon is delegated to an individual responsible for its implementation to an agreed timescale. Corrective action procedures also form part of the internal auditing and management review procedures (see Figure 2.5).

Sub-clause 4.15: Handling, storage, packaging and delivery

Procedures should be established to provide an effective means of preserving and marketing products and materials:

1. *Handling*: The methods and equipment used for the purpose of product handling should prevent damage and deterioration
2. *Storage*: Suitable, designated areas are required for the protection and control of products and materials which are to be used in the manufacturing process or are to be delivered to the customer
3. *Packaging*: The packaging instructions must be in conformance with specified requirements. Relevant documentation may consider:

- Customer complaints
- Quality plans
- Route cards
- Quality control records (inspection and testing)
- Calibration records
- Allowed concessions
- Process equipment operating records
- Works instructions (process and installation)
- Works order documents including drawings and specifications

Figure 2.5 *Possible sources of associated documentation for action*

- Methods of cleansing
- Methods of preservation
- Details of packaging
- Method of identification.

Company responsibility may also extend contractually to delivery of finished goods to a client. Effective storage must be provided if this is the case.

Sub-clause 4.16: Quality records

As the complete quality system is documented in quality and procedures manuals, the proof of effectiveness is to be found in the records. There must be provision of evidence of quality through records of orders; purchases; complaints; rejections; audits, etc. These documents (verification documents) must be retained in a controlled filing system for a given period determined by internal criteria and contractual stipulation (for example, five years for product liability purposes). A register of all quality records must be kept by the management representative for quality and should include:

- Internal audit reports
- External audit reports
- Management review records
- Contract review records
- Procurement records
- Process control documents
- Material certification
- Inspection and test data
- Calibration results
- Statistical methods results
- Customer/supplier complaints

- Supplier approved ratings
- Non-conformance, corrective action and concession reports
- Training records

Sub-clause 4.17: Internal quality audit

It is essential that there is a procedure for periodic checking of th performance of internal audit. The formalized procedure will enabl management to assess efficiency and identify any weakness in th quality system or associated training. Each element of the standar should be audited each year or as circumstances dictate (i.e. change to specifications, deficiencies identified, etc.), by personnel skilled i auditing procedures. Auditors should be appointed, wherever possible from an area independent of that being audited. The results of eac audit must be recorded in audit, non-compliance and corrective actio reports. A formal audit review meeting will communicate the results t the personnel responsible for the area under assessment.

Sub-clause 4.18: Training

The success of the quality system depends on effective training o personnel. A training programme should consider:

- Quality awareness and understanding
- Significance of quality
- Quality allied to competitiveness of company
- Individual role in company context
- Job role knowledge and skills
- Ongoing review of training needs.

Ongoing identification, provision and review of training allows fo knowledge and skill levels to be assessed against possible development within activities of the quality system (see Chapter 9).

Sub-clause 4.19: Servicing

The servicing of a product, post-delivery and installation, may constitute an integral part of an overall contract. Servicing requirements should be incorporated where applicable at the contract review stage This enables a servicing quality plan incorporating product specification, product drawings and servicing manual to be drawn up, which indicates the sequence of events in a particular servicing activity. The personnel allocated to servicing activities must be suitably trained and documentation relevant to their activities (service reports) must provide evidence of having complied with service requirements.

Sub-clause 4.20: Statistical techniques

The use of statistical techniques can be an effective method of evaluating the level of quality capability of a given process. They may be used because of contractual requirements or as a result of internal company policy. The measurement of variation within product and process provides the data for statistical assessment of quality levels (actual and potential). The use of statistical methods is not limited to past activity and can be used for the following purposes:

- Market analysis
- Product design
- Process control studies
- Data analysis
- Performance assessment
- Determining quality levels
- Determining inspection items
- Reliability specification

Specific statistical quality control methods include:

- Frequency distributions
- Control charts
- Regression analysis
- Measures of central tendency
- Acceptance sampling
- Risk analysis
- Safety evaluation

Quality system assessments and audits

An assessment of an organization's quality system is a formal appraisal of that system and is carried out to establish to what extent the system meets the criteria specified in a quality standard or a quality assurance scheme (ISO 9000 is one such quality system standard). The assessment is conducted by a body which is certified under the National Accreditation Scheme. It involves a documentation review and a company visit to determine if the quality management system:

- Exists
- Is correctly operated and maintained
- Is effective.

First, the quality system must exist in a documented form, typically a quality manual and operating procedures. Second, there must be evidence in terms of records that the system has been implemented and is

being operated correctly and according to the documented procedures
Third, the quality system must be effective in terms of:

- Early detection of defective material
- Analysis of quality problems
- Improvements made in existing processes, products
- Benefits being identified.

During the assessment, the assessor seeks evidence of:

- Clearly defined responsibilities and authority
- Documented procedures, instructions and controls
- Knowledge and understanding of responsibilities, authority, pro- cedures and instructions
- Correct operation of procedures by the authorized and responsibl personnel
- Adequacy of personnel, equipment, facilities and general resources
- Effectiveness of the system when correctly operated.

ISO 9000 requires a management system which will ensure that th customer's specification requirements are met. A management repre sentative has to be appointed who shall have defined authority anc responsibility for ensuring that the requirements of the standard ar implemented and maintained. In other words, there has to be a foca point for the whole quality system; somebody who will manage anc coordinate the system on behalf of the company (see Figure 2.6)

1st party	Assessment by company itself to ensure compliance with requisite standards for a contract.
2nd party	Intermediate assessment by a major purchasing body of suppliers, policies and procedures to ensure conformity to an accepted standard.
3rd party	Assessment carried out by an independent certifying body which does not itself trade with the company. Assessment of applicant's quality management system and subsequent rechecking of production and finished products to ensure that they continue to meet the necessary standards.

Figure 2.6 *Certification schemes*

Quality auditing

If an organization wishes to become accredited to a quality system standard such as ISO 9000 or is already approved to a quality system standard such as ISO 9000 and requires a periodic check to establish continuing compliance, it must undertake to carry out an *audit*. The quality assurance audit is an objective evaluation of the effectiveness of the quality programme and its component parts. It is a procedure to verify facts, the purpose of which is not to control but to confirm whether or not we have control. The true quality audit provides an index which can be used as a measure of the quality system, and as a measure of quality and its shifts and trends with time. An audit can then provide information that will aid in taking intelligent action. British Standard 4778 (ISO 8402) defines a quality audit as:

> A systematic and independent examination to determine whether quality activities and related results comply with planned arrangements and whether these arrangements are implemented effectively and are suitable to achieve objectives.[3]

A well-planned audit programme should result in management action. Such action may result in modifying policies, systems and procedures to ensure that a desirable level of product or service quality is achieved with economic use of resources. Auditing is a discipline devoted to the establishment of facts. It is not an inquisition of people. Audits must be planned and executed competently and objectively. They must be reported constructively to achieve maximum improvements in the total quality programme. The audit conclusions must be communicated to an appropriate level of management for action. Objectivity can best be achieved if the group responsible for checking, inspecting or otherwise verifying that an activity has been correctly performed, is independent of the group directly responsible for performing the specific activity.

Purpose of auditing

Several specific purposes of internal audits are given below. When such audits are adequately supported and properly conducted, the results will provide valuable data to:

- Assess the effectiveness of quality assurance activities and operations throughout the company by means of a planned audit programme
- Ensure compliance with company quality policies, systems and procedures
- Measure the degree of effectiveness of quality systems
- Evaluate the effectiveness of people in the implementation of quality plans
- Optimize quality/cost relationship
- Identify quality weakness that might result in a crisis
- Provide information for major changes and accelerate the evolutionary process of quality improvement
- Promote understanding between different departments

- Contribute to technology transfer
- Communicate to management
- Reduce customer complaints.

Types of audits

The subject matter of quality audits extends across the entire spectrun of the quality function, but the bulk of the auditing is carried out unde the following categories:

1 *Audit of policies and objectives*: The scope of this audit extends to th business aspects of the company's quality activities as well as th technological aspects. The standard used to judge the adequacy o quality policies and objectives is a mixture of past performance o competitors and of subjective judgement
2 *Audit of performance against company objectives*: Normally, a reviev based largely on the data presented by the executive reports o quality. The reviews are conducted by upper management, th usual frequencies being quarterly or monthly
3 *Systems audit*: This is an audit of the quality manual to ensure tha it meets the requirements of the ISO 9000 series of standards
4 *Compliance audit*: This is an audit to determine if the implementa tion follows the procedures, work instructions, quality plans, etc.
5 *Product audit*: This determines if the product meets the specifica tions and the needs of fitness for use.

Conducting a systems audit

The auditor (the individual authorized to conduct internal audits accepts the base system as being 'approved' and checks that it i complied with throughout:

1 Confirm that the quality manual and issue status are pertinent tc the audit, e.g. it is the issue which has been approved by a third party accrediting body
2 Establish if the quality manual and other documents which pre scribe the quality system are up to date
3 Compile a suitable questionnaire, or if the one used for the assess ment is available, use that one
4 Confirm that the organization and assignment of responsibilities are as described. Check that each post is filled and the departmental staff is up to complement
5 Visit each technical office and production department to verify that the practices comply with the system prescribed. Question person nel and examine sample documents. Seek corroborative evidence Use all possible techniques to establish corroboration

6 Base the audit on the written statements in the approved documents. Do not refer to 'good practice' or to your own preferences
7 Determine whether or not the manufacturer's practices throughout his or her offices and shops comply with his approved quality system
8 When reporting the audit, list the elements of the quality system which have been found to be satisfactory. For the elements which have been found to be unsatisfactory, report factually and in detail the reasons why they are.

Techniques used in assessments and audits

The assessment and audit techniques are the methods used to find the necessary information. They include:

1. Reviewing the quality manual
2. Reviewing departmental procedures
3. Asking direct questions
4. Asking indirect questions
5. Examining documents which verify statements made
6. Examining one document to verify the information given on another
7. Examining documents which specify the procedures to be followed and checking compliance with the procedures
8. Comparing the spoken word with the written one.

Wherever possible, seek corroborative evidence of any statement made or written.

(a) When a person makes a statement, ask a suitable question of another which will confirm the answer given by the first person:
e.g. To person 'A': What do *you* do?
Why?
To person 'B': What does *he* do?
Why?

(b) When a person makes a statement, whenever possible, examine documents to verify the statement:
Example 1.
To a departmental manager: What are *your* responsibilities?
To verify the answer: Examine the quality manual.
Examine the depatmental manager's job description.
Example 2.
To a buyer: How do you order materials?
To verify the answer: Examine the quality manual.
Examine the purchasing department procedures.

Check a purchase requisition.
Check a copy of the order.
Check data on the order.
Check the list of approved suppliers.
Check the acknowledgement of the order.
Check and follow through any amendments to the order.

Example 3.
To a welder: Which weld procedure are you using?
To verify the answer: Examine the welder's job instructions.
Examine the weld procedure.
Check that the issue is pertinent.
Examine the requisition made out by the welder for consumables.
Examine the stores record of consumables issued.
Make physical checks on the set-up, consumables, equipment, and the current.
Compare these with the procedure.

Example 4.
To a tester: How do you carry out the test?
To verify the answer: Examine the tester's job instructions.
Examine the test procedure.
Check the test procedure against the requirements of the code.
Make physical checks on the set-up, equipment, calibration and readings.

Whenever possible, try to establish a chain reaction. Take one contract and check the information through the chain.

Close loop

On important matters, look for a 'closed loop' system. That is a system by which a person who decides what has to be done is later advised that it has been done.

Example 1: The sender of an important document receives an acknowledgement of receipt from the addressee.
Example 2: The purchaser who prescribes what the finished dimensions should be, receives an 'as-built' report.
Documents: For each important document, there must be formal procedures for:

1 Writing
2 Content
3 Review

4 Approval
5 Identification
6 Issue status
7 A controlled distribution
8 Changes
9 Approval of changes
10 Issue revision
11 Re-distribution

Enquire for each document if there is a procedure which specifies how the above operations should be carried out.
Example 1:
Check: Take a sample works order and check that it has been prepared, approved and distributed in accordance with the prescribed procedure. Review the client's order. Note any amendments. Trace the amendments through the works order and redistribution. Check that the correct procedure has been followed and the correct records kept.
Example 2:
Check: Take a sample drawing from the shopfloor.
Note: The number and issue status.
Visit: The drawing office. Check the original and the records of compilation, approval, distribution. Check that the drawing has been prepared, approved, identified in accordance with the correct procedure.
Alterations: Examine the alterations to the drawing. Check the related records – e.g. change notices, redistribution notices. Check that all these have been prepared, approved, identified and redistributed in the prescribed manner.
What? Who? Why?
For each document ask:

What is its purpose?
Who should see it?
Why?

Wherever possible, relate your question to the answer to the previous question.

Benefits of accreditation to ISO 9000

There is a growing tendency in industry and commerce world-wide to implement formal quality systems such as ISO 9000. For example, in Britain some 20 000 organizations have, or are in the process of seeking, accreditation for the ISO 9000 standard. The reasons for this current level of activity are many. Customer requirements, National and European requirements for inclusion on tender lists, and the harmonization of standards, all play a part. However, there is a danger that organizations may adopt a formal quality management system fo operational reasons only, often as a reaction to changing competitiv conditions. As a consequence of such an operational rather than stra tegic focus, the benefits accruing from successful implementations ma not be attained.

There are three basic reasons why companies may choose to g through the ISO 9000 registration process:

- For the intrinsic value gained from meeting the challenge
- To meet the requirements of a single large supplier such as som government agencies that are requiring ISO 9000 or equivalent com pliance for large contracts
- To maintain or gain access to markets, particularly in Europe, where quality plays a special role in the emerging set of harmonized product safety standards.

Fundamentally, third-party certification under the ISO 9000 standard means that a company can demonstrate that it has an appropriat quality system for the products or services it offers. Moreover, standar dized certification means that a single audit process can assure pur chaser organizations that a supplier company's quality system meet requirements, and redundant audits by several of a company's custo mers can be eliminated or at least scaled down.

The implementation of an ISO 9000 quality system has potentia advantages:

1 Having an organized form of communication means:
   - Improved management
   - Better planning of all activities
   - Early resolution of problems
2 More precise specifications means:
   - Correct interpretation of customer needs
   - Better chance of complying
   - Identification of weaknesses in specifications/orders
3 Greater control of sub-contractors and suppliers
4 Increased efficiency giving a better-quality product at no extra cos and increased productivity
5 Less remedial work and scrap
6 Feedback of customer problems and more rapid correction of inad- equate production methods
7 Improved performance in meeting target delivery dates
8 General increase in the standard of workmanship and therefore a more satisfied customer
9 Improvement of the reputation of the manufacturer.

Various sources confirm that the actual benefits to be accrued from certification relate to the perceived advantages:

1 As a result of the implementation of an ISO standard, companies experience:
   - A decrease in the number of customer complaints
   - An increase in the number of sales
   - An increase in the number of repetitive customers
   - An increase in the number of satisfied customers
2 The standard helps to highlight troublesome areas in the production process and enables personnel to gain a more thorough knowledge of processes
3 Internal audits prove to be an effective way of highlighting deficiencies in the whole system in that they reveal areas for improvement and, significantly, assess the efficiency or inefficiency of management
4 The documentation process aids the training and development of employees
5 Better use is made of scarce resources with the result being
   - Lower defect rates
   - More individual responsibility for quality
   - Less waste
   - More focus on problem solving.

One of the potential advantages of pursuing accreditation to ISO 9000 is that there is increased efficiency giving a better-quality product at no extra cost. Ultimately, registration to ISO 9000 may be cost efficient in that the continued focus on process improvements results in return on investment. However, in the short term it is necessary to be aware of the cost implications of gaining and maintaining accreditation to ISO 9000.

In terms of gaining certification, the cost element can be split into two categories. One is the internal time and effort required to design and install the system. The other consists of the external costs which include certification fees, consultancy fees, the costs, of new equipment or services, the costs of reorganization and training costs.

The issue of quality costs is one to which the ISO 9000 system standards do not directly refer. There is no specific requirement to monitor quality costs. Most organizations who pursue accreditation regard the prevention of waste and mistakes as a fundamental task of management, which could not be separately costed.

The subject of quality cost models is under review and an addition to the British Standard, BS 6143, is currently being prepared to incorporate new approaches to quality costing.[4] Whether or not the inclusion of a quality cost element becomes a mandatory element for the ISO 9000 standards, organizations should attempt to progressively eliminate costs which do not add value to the product, such as inspection and test, internal transportation and storage. Quality costs will be addressed in Chapter 6.

Planning the quality system

When the decision is made by an organization to implement a qualit system, a manager should be appointed who is to be responsible fo planning and implementation of the quality management system. Th position may be full- or part-time, depending on the complexities of th organization's activities and the extent to which it has already focuse on quality improvement activity.

Managerial responsibilities

1 *Review requirements*: Having determined the quality managemen system to be implemented, senior management and the qualit assurance manager (project manager) should familiarize them selves with the requirements of the system
2 *Prepare or revise organization charts*: These should indicate areas o individual managerial responsibilities and the lines of account ability
3 *Prepare job descriptions*: Managerial responsibilities are clarified i the form of a job description for each manager. These should high light specific responsibilities for product quality as well as relate t other managerial responsibilities
4 *Form a project management team*: Other managers and employee should contribute to the planning of the system. The project man ager should control these related activities
5 *Prepare working procedures*: All aspects of the company's opera tions which affect the quality of product should be documente as procedures or work instructions. The procedures should be writ ten by those managers responsible for the activities relating to th activities.

Job descriptions

Customer services manager

Reports to financial controller and is responsible for:

1 Controlling receipt and acknowledgement of all orders
2 Interface with customers on all queries and information regarding status of orders
3 Quotations for specials and new products
4 Interface with shippers and customs officials on matters relating to shipments
5 Liaison with engineering on new product schedules and/or changes
6 Creating bills of material and documentation for 'specials'

7 Acting as principal interface with the sales and customer services organization
8 Booking and scheduling orders to meet monthly schedule.

Production services supervisor

Reports to the general manager and is responsible for:

1 All aspects of quality assurance within the company:
   (a) Definition of procedures
   (b) Assessment of processes
   (c) Identification of documentation needs as impacts upon quality
   (d) Evaluation of application procedures.
2 Day-to-day operation of quality function in regard to incoming, in-process and finished goods.
3 Control of warranty and customer complaints.
4 Control of master drawings:
   (a) Receipt from draughting
   (b) Issue to points of use
   (c) Obsolescence of 'old' drawings.
5 Control of documentation:
   (a) Change control
   (b) New component approval
   (c) Corrective action requests.
6 Approval of vendors/sub-contractors in liaison with purchasing.
7 Assessment of vendor quality systems where necessary.
8 Control and calibration of inspection, measurement and test equipment.
9 Supervision and direction of production engineering.
10 Control of stores:
   (a) Receipts
   (b) Issues
   (c) Stock checks
   (d) Storages
   (e) Obsolescence.
11 Act as management representative with responsibility for the implementation and maintenance of requirements of quality management systems.

Quality/training supervisor

Reports to the production services supervisor and is responsible for:

1 Devising and implementing training plans for all new employees. Retraining existing employees in new tasks, and carrying out these plans

2 Directing and supervising inspection personnel on a day-to-da basis
3 Liaising with engineering during the introduction of new product and introducing these new products into production with the assis tance of the production engineer
4 Acting as in-house safety officer. Carrying out safety audits an monitoring staff adherence to safety regulations/guidelines
5 Deputise for production services supervisor as head of qualit when necessary.

How to implement the quality system

1 *Determine priorities*: As it is not feasible to implement all the rele vant procedures at once, they should be reviewed by senior man agement and the project manager to ensure accuracy and to decid upon the method of implementation
2 *Identify key personnel*: Certain employees will be of benefit to th planning and implementation of the system in that they can hel identify quality problems such as:
   - Troublesome products
   - Poor maintenance
   - Lack of training
   - Inadequate specifications, drawings, etc.
   - Lack of time, etc.
3 *Communication of information*:
   - Dates for introduction of procedures
   - Training information relevant to quality system
   - Timescales for related activities
   - Personnel responsible
4 *Generate commitment*: If senior management are not fully com mitted their diffidence will be apparent to other employees an the project will either fail or not produce the desired results Demonstration of commitment is essential before implementation
5 *Demonstrate progress*: When implementation begins, it is importan to demonstrate practical success. A well-planned system will enabl information to be collected and analysed relevant to customer satis faction, product characteristics, purchaser, supplier, product, etc This information will form the basis for improvement activity regarding the determination and elimination of defects.

Service quality

The quality assurance standards might, at first glance, appear to be solely applicable to manufactured items, and service companies would find them difficult to apply. However, many of the criteria of a quality programme (i.e. BS 5750, ISO 9000) also relate to service industries. For example, every company, regardless of the industry in which it operates, would require:

- Quality programme
- Control of purchased material and services
- Quality organization
- Records
- Audits
- Non-conformances
- Quality programme documents
- Corrective action
- Planning
- Training
- Documentation control.

Airlines, banks, hotels and many other kinds of service organizations have highly developed quality control systems. However, these systems are generally based on industrial quality assurance systems and tend to be product oriented rather than service oriented. They are designed to:

- Ensure properly made-up hotel rooms
- Clean rental cars
- On-time landings
- Banking transactions with a minimum of clerical errors.

In other words, they deal with the technical aspects of providing a service and not with the total service from the customer's point of view. For example, the technical aspects of providing service in a bank include:

- Time to clear a cheque
- Availability of foreign currency
- Renewal of cheque book
- Money available at cash dispenser.

The status of ISO 9000 as it applies to services and the services sector has been addressed by the publication of the ISO series Guidelines for Services (ISO 9004, Part 2) in draft form in 1990. Although an organization cannot become registered to ISO 9004, it provides a useful mechanism for dealing with the internal as well as external customer. In the majority of companies the internal customer constitutes up to 80% of the market as the percentage of employees actually making things is approximately 20%. The rest is all service.

The scope of the standard covers the definition of 'service'. Descriptive characteristics include how much product content there is and how long the service lasts. Car or audio equipment repairs, for

example, have a high product content, while the services of a solicito
have a low one. Service duration may vary from the length of
phone call as when a public telephone is used to an extended perio
covering weeks or months.

One of the key elements of the ISO 9004-2 standard is the require
ment that a company's service be defined with specific characteristic
documented such as dependability, capacity, safety, security, courtes
and accuracy. The standard also addresses the importance of employe
involvement and motivation in providing quality service, and vesting
service quality loop which enables internal and external measuremen
of customer satisfaction (see Figure 2.7).

Quality system audit checklist

The purpose of this guideline is to ensure uniformity in the preparatio
of the audit checklist and standardization of audits performed in suc
cessive years. The intent is to provide an audit baseline while no
limiting the auditor from investigating additional areas when appro
priate. Analysis of results of succeeding audits can then be made wit
the knowledge that the audits performed were comparable.

The checklist provides recommendations for quality system audits
The quality system has been separated into key sub-systems which
support the overall effectiveness and provides questions to assist i
the evaluation of each sub-system. Since products, facilities and man
agement techniques vary so widely, it is not practical to generat
detailed checklists for all levels of suppliers. The following checklis
is a database that should be modified specifically towards the specia
requirements of a company.

1 Management responsibilities
  (a) Is there a documented quality policy?
  (b) Has the quality policy been communicated to relevant personnel?
  (c) Are quality objectives documented?
  (d) Do these objectives pertain to key elements of the performance reliability and safety of the service?
  (e) Are personnel responsibilities for quality defined and documented?
  (f) Is there an independent review and evaluation carried out of the continuing stability and effectiveness of the quality system?
2 Review of customer requirements
  (a) Is there an internal customer focus within your organization?
  (b) How does your organization continually keep in touch with the needs of the internal customer?

Administration	**Personnel**
Typing throughout time	Cost per recruitment
Documenting error rate	Time taken to fill vacancies
Proportion of documents queried	Training needs unfulfilled
Customer complaints rate	
Debtors and creditors outstanding against schedule	

Reliability	Consistency of programme and dependability	• Accuracy in invoicing • Keeping records completely
Responsiveness	Willingness of employee to provide service	• Calling customers back quickly • Giving prompt service
Competence	Having required skills and knowledge to provide service	• Knowledge and skill of personnel
Courtesy	Politeness, respect, consideration	• Consideration for consumer • Neat appearance of contact personnel
Communication	Keeping customers informed	• Explaining service • Explaining cost of service • Assuming customer problems will be dealt with
Credibility	Trustworthiness, honesty	• Company reputation • Personal characteristics of contact personnel
Understanding/ knowing the customer	Understanding customers' needs	• Learning customers' specific requirements • Providing individualized attention
Tangibles	Physical evidence of service	• Physical facilities • Appearance of personnel • Tools or equipment used to provide service

Figure 2.7 *Examples of performance measures used in service areas and determinants of service quality*

- (c) Is there a quality review of purchase orders to identify special or unusual requirements?
- (d) Are requirements for special controls, facilities and skills pre-planned to ensure that they will be in place when needed?
- (e) Are customer requirements available to personnel involved in the manufacture, control and inspection of the product?
- (f) Are supplier sketches, drawings and specifications compatible with the customer's requirements?

3 Supplier control practices

- (a) Are your suppliers aware of what your quality requirements and expectations are?
- (b) Is there a system for identifying qualified sources and is this system adhered to by the purchasing function?
- (c) Are initial audits of major suppliers conducted?
- (d) Does the system ensure that technical data (drawings, specifications, etc.) are included in purchase orders?
- (e) Are the number and frequency of inspections and tests adjusted based on supplier performance?

4 Documentation

- (a) Is there a quality manual documented?
- (b) How is the quality manual updated?
- (c) Is there a documented quality plan for each project (e.g. sequence of events, accountability)?

5 Manufacturing and quality planning

- (a) Does the quality planning function account for all product and process characteristics? The following should be considered:
  - Process capability
  - Potential process-generated imperfections
  - Characteristics that change during manufacturing sequence
  - Evaluating characteristics as close to their generation as possible
  - Items with performance requirements
  - Customer's quality level requirements
- (b) Is suitable measurement equipment used?
- (c) What system of control is used (operator/assembler acceptance, in-process/roving inspection, tollgate inspection, end-of-line inspection, process control, etc.)?
- (d) Is the sequence of manufacturing and inspection operations specified?
- (e) Do the procedures specify sample size, frequency, gauges to be used, and special methods?
- (f) Are the requirements for quality measurements specified? Do they include who makes the measurements, which measurements need to be recorded, and whether recording is variable or attribute?

(g) Is manufacturing and quality planning current?

6 Material identification/routing practices
   (a) Does the sequence of manufacturing and inspection operations (routing sheets/travellers) properly identify the parts for which they apply?
   (b) Are lots identified in a manner traceable to customer purchase orders?
   (c) Is there a means to associate specific parts with their routing sheets/travellers?

7 Non-conforming material
   (a) Are non-conformances identified and documented?
   (b) Are non-conformances physically segregated from conforming material where practical?
   (c) Is further processing of non-conforming items restricted until an authorized disposition is received?
   (d) Do suppliers know how to handle conformances?
   (e) Are process capability studies used as a part of the non-conforming material control and process planning?

8 Design and process change control routines
   (a) Are changes initiated by customers incorporated as specified?
   (b) Are internally initiated changes in processing reviewed to determine if they require customer approval?
   (c) Is the introduction date of changes documented?
   (d) Is there a method of notifying sub-tier suppliers of applicable changes?
   (e) Are periodic design reviews carried out?
   (f) Are new and modified services validated prior to implementation?

9 Material handling and housekeeping practices
   (a) Is housekeeping adequate?
   (b) Are parts adequately protected during handling?
   (c) Are suitable containers provided for the shape and size of in-process and finished parts?
   (d) Are openings in parts covered to prevent potential entry of dirt, chips and contamination?
   (e) When solutions and powders are used, are special precautions used to prevent their residual deposits in openings and on critical surfaces?

10 Quality records and retention
   (a) Are records maintained that show compliance with work requirements?
   (b) Do inspection/test records show the nature and number of observations made, the number and type of deficiencies found, the quantities accepted and rejected, and the disposition of non-conforming items?

(c) Are effective records kept to analyse by process, people, par number and operations?
(d) Does management analyse and use these records as a basis fo decisions?
(e) Are records retained for the period specified by internal o customer requirements?

11 Control and maintenance of quality measurement equipment
(a) Are calibration methods documented?
(b) Is a positive recall system in place to ensure that out-of-cycl tools, gauges and instruments are not in use?
(c) Are calibration intervals established and adjusted based o stability, purpose, degree of usage and previous calibratio results?
(d) Are all tools, gauges and instruments used for product accep tance included in the calibration system?

12 Corrective action
(a) Are non-conformances reviewed to determine root cause?
(b) Is corrective action implemented when required?
(c) Are corrective actions reviewed after implementation to verif their effectiveness?
(d) Does the corrective action system include deficiencies in pro cesses, methods, facilities and systems?
(e) Are customers notified when an investigation reveals that non conforming material may have been delivered?

13 Process and products audits
(a) Are process audits conducted?
(b) Are product audits used independent of normal product accep tance plans?
(c) Do the audits cover all operations, shifts and products?
(d) Do audit results receive management review?
(e) Is the audit frequency adjusted based on observed trends?

14 System audits
(a) Are system audits conducted?
(b) Are all sections and shifts of the quality system audited?
(c) Are audit results and corrective action results reviewed with the company's management?

15 Personnel skills and communication
(a) Are training needs identified?
(b) Is there a training plan?
(c) How is training carried out?
(d) What training in quality is provided?
(e) What are the methods of communicating with personnel withir the company?
(f) Are formal meetings held with employees?

eyond quality systems

Quality management is inextricably linked with the overall management of a business and ISO 9000 may be used as a basic tool to improve its efficiency and effectiveness. Some companies do attempt to extend the principles of quality management systems such as ISO 9000 throughout the company, while others adopt a more jaundiced view, considering ISO 9000 as a method of documenting their usual business practices which lack quality.

Too many organizations fail to realize the strategic importance of quality. ISO 9000 still constitutes vertical thinking in that it promotes accountability of the process but does not impinge on all those business activities which determine the capability of an organization to satisfy customer requirements.

The potential benefits to be accrued from implementation of a quality measurement system such as ISO 9000 are rarely viewed in strategic terms. Yet the role of ISO 9000 in process management is an important indicator, among others, of how an organization systematically manages its key and support processes.

The management review clause of ISO 9000 has particular relevance in this context in that it can be used as both a communications forum and as an opportunity to appraise the suitability of all business solutions, not just quality tasks. The language of the standards, however, results in common interpretation of the remit which does not extend beyond the production-related activities into the service functions, (such as marketing, accounts, administration, etc.). There is also an implication of continuous quality improvement in the corrective action clause which requires a company to investigate, analyse, prevent non-conformities and record action to deal with non-conforming products.

The misconception that ISO 9000 certification can guarantee a quality product or service mirrors that which perceives effective management of processes to be an all-encompassing approach to quality improvement. Effective management of processes using a system such as ISO leads to consistency. A customer could, in fact, receive either an excellent or a poor service consistently, depending on the level of commitment and the effectiveness of the supplier's quality system. However, as indicated above, there are other aspects of organizational activity to be considered – policy and strategy, people management, people satisfaction, quality costs, etc. The lack of commonality in respect of a definition of quality/total quality stems from the varied interpretations held by different organizations. To some, it is an approach that makes customer satisfaction the primary focus, whereas others place emphasis on problem-solving teams or statistical process control.

The British Standards Institution published a standard in 1992 – BS 7850 Total Quality Management.[5] While it is not a standard to which companies will be able to register, it is intended to be used as a guide to help companies who are embarking upon a total quality continuous improvement initiative or are wishing to assess progress made. The standard is based on the following definition of TQM:

> The management philosophy and company practices that aim to harness the human and material resources of an organization in the most effective way to achieve the objectives of the organization.

There is a recognition that TQM requires top-level commitment and involvement:

Part 1 : *Guide to Management Principles*: This part is designed to help senior management identify how to establish the management principles that will maximize effectiveness in meeting organizational objectives

Part 2 : *Guide to Quality Improvement Methods*: This part deals with the implementation of a continuous quality improvement process as applied to every aspect of an organization. The most commonly used tools and techniques are also highlighted.

Further clarification of one definition of TQM has been facilitated by the development of the European Quality Award. The award criterion defines a total quality organization as one that moves beyond short-term financial goals when determining its capabilities. Such an organization has processes or methodologies driving its results. These processes are continually reviewed for effectiveness and for appropriateness to changing situations. A commitment to continuous improvement should also exist and future performance benchmarked against competitors rather than previous achievement. Such a level of performance requires a high rate of change which can only be achieved through using and developing the capabilities of all employees. Success is also dependent on the development of a strategy and associated plans which are based on customer satisfaction.

Although this model forms the basis of the European Quality Award, the main purpose is for company self-appraisal. The process of self-appraisal consists of the regular and systematic assessment of an organization's activities and results against best practice. It enables the organization to clearly identify strengths and areas for improvement by using each criterion of the model.

All organizations are unique. However, the model provides a framework for self-appraisal which is applicable to virtually every business organization. It is more comprehensive than the traditional quality audit in that the entire activities of the organizations are assessed rather than merely the quality system.

Summary points

- The quality system of an organization is characterized by its network of processes and the responsibilities, procedures and resources that relate to the processes.
- BS 5750/ISO 9000 provides a framework for developing a quality system. It sets out the method by which a management system, incorporating all the activities associated with quality, can be implemented in an organization to ensure that all special preformance requirements and the needs of the customer are fully met.
- BS 5750/ISO 9000 can be used as a framework for designing your own system.
- The implementation of ISO 9000 has many potential advantages including:
  (i) better planning of activities;
  (ii) correct interpretation of customer needs;
  (iii) early resolution of problems;
  (iv) less remedial work and scrap; and
  (v) feedback of customer problems.
- A well-planned audit programme should result in management action.
- BS 5750/ISO 9000 is an important indicator of how an organization systematically manages its processes.
- BS 5750/ISO 9000 does not impinge on all those business activities which determine the capability of an organization to satisfy customer requirements.

References

1 BS 5179, Guide to the Operation and Evaluation of Quality Assurance Systems, British Standards Institution. London.
2 BS 5750/ISO 9000, Quality Management Systems, British Standards Institution, London, 1987.
3 BS 4778: Part 1: 1986, Quality Vocabulary: International Terms, British Standards Institution, London.
4 BS 6143: Part 2: 1990, Guide to the Economics of Quality, British Standards Institution, London.
5 BS 7850: 1992, Total Quality Management, British Standards Institution, London.

3 Determining the capability for quality

Introduction

The importance of the customer in the management of quality has been stressed throughout this text. It is clear that the customer determines the level of quality and not the supplier. If this is accepted then there is an additional 'burden' put on the supplying organization. If that organization wishes to aim for world-class status and satisfy/delight its customers, then it must understand two key issues:

1 What the customer requirements are
2 What their own capabilities are.

By understanding both these issues an organization is in a position to identify the gap between what the customer needs and what the organization can provide. This alone should stimulate improvement activity.

This chapter seeks to outline the key issues involved in using customer and indeed supplier feedback to promote improvement and allow effective target setting. Its objectives are:

- To develop an understanding of the relationship between customer, supplier and organization
- To discuss how to identify, gather and analyse data which will help initiate and facilitate improvement activity
- To recognize the benefits of a systematic approach to the identification of critical processes
- To reveal how to manage, review and improve processes
- To develop awareness that the effective management of processes is an important indication of organization performance.

What is a process?

The principles of quality improvement derive from simple definitions:

- A product or service is any and all output perceived by a customer
- A *process* consists of all the activities carried out to generate the output, including controls such as procedures and resources such as people
- A customer is the recipient of the output

- A supplier is the provider of the output.

A process, then, is all the internal operations and sequences of operations used to generate and deliver a produce or service. These can include design, competitive analysis, scheduling, planning, engineering, production, accounts payable etc.

If a process can be defined as the grouping in sequence of all those tasks directed at accomplishing one particular outcome, then every activity becomes part of a process.

Input → Process → Output

A process focus results in work being considered as a series of related and interrelated tasks. Thus:

- The relationships between jobs become apparent – individuals are customers of each other in an organizational context, irrespective of status
- Employees are able to develop a unified language and an understanding of what their jobs entail
- An improved understanding of activity enables employees to define the capacity of existing processes to satisfy customer requirements
- Employees will be able to focus on errors, waste and other problems in a preventative mind set.

Identifying critical processes

Process management is based on the premise that all work carried out in an organization is interrelated and part of some overall business process. Business processes are groups of logically related tasks that utilize the resources of the organization to provide results required to achieve the business objectives. Processes are defined independently of organizational structure. A set of activities comprising a business process may exist within one organizational unit or it may cross organizational boundaries. Also, tasks that an individual performs may constitute activity elements within a number of different processes. All organizations have a large number of processes which contribute to the provision of a product or service.

Furthermore, the identification of improvement opportunities occurs constantly everywhere and in relation to those processes. It is necessary, therefore, to identify those processes which are critical to the success of the organization. Critical processes could include design for manufacture, design of service, determination of people satisfaction, etc. The basis for the selection of critical business processes are the strategies of the organization which are usually formulated in the organization's strategic plan and other policy documents. The mai strategies provide the image an organization wants to portray in it market or environment. This image is then condensed to the essentia activities and their respective targets. The selection of critical processe can be facilitated by considering its relative importance on three level

1 On the organization's products, services and information
2 On the impact of the process:
   - On usage
   - On the environment
   - If it were to be withdrawn
3 On the resources needed to perform the activities involved in th process.

Exercise 3.1

Using the above criteria, identify the critical processes of your organ ization.

Process management

The management of processes is based on two preconditions:

1 The knowledge of customer requirements and of their expectations
2 The expertise to analyse the processes for improvement opportu nities as well as to implement improvements in the process.

The analysis basically consists of:

1 The identification of the process including the sub-processes. Fo example, if 'recruiting personnel' is a process, its sub-process migh include:
   - Identification of the need
   - Preparation of candidate profile
   - Preparation of job description
   - Identification
   - How to publicize a job
   - Preparing contacts
   - Interview
   - Decision on offer

2 The formulation of critical success factors (CSF), i.e. key indicator or prerequisites of the required process output performance. If th required output from the recruitment process is an effective mem ber of staff, critical success factors might include:

- Accessibility to sources
- Knowledge of candidate's requirements
- Administrative procedure
- Participation in preparing the job description
- Attractiveness of offer

3 Evaluating the efficiency of the most important processes for the selection of 'critical sub-processes' which can become the focus of improvement activity.

The key measurable attributes of any process are efficiency and effectiveness, where efficiency is 'doing things right' and effectiveness is 'doing the right things'. A process may be efficient in that it follows a logical sequence of predetermined activities while operating with minimum use of resources, but it may not be effective in actually meeting customer requirements, as certain steps in the sequence of activities may not be adding value to the process. Examples of measurements could include:

1 Effectiveness-oriented measures, i.e. error rate, response time, accuracy, etc.
2 Efficiency-oriented measures, i.e. turnaround time, asset utilization, time per activity.

Process improvement

There are three basic elements that need to be addressed when an organization positions itself for continuous improvement. These are:

1 Determining exactly what the organization's business objectives are, based on a customer focus
2 Analysing the organization's processes to determine and track areas for improvement
3 Solving problems.

They are referred to as departmental task analysis, business process analysis, and problem solving.

Departmental task analysis

An important task in pursuing quality improvement is for the members of every organization to determine exactly what its mission and responsibilities are and how it interacts with customers and suppliers. The process is known as departmental task analysis (DTA). DTA typicall begins by bringing members of the organization together in a locatio where they will not be disturbed. There, they write a mission statemen determine organizational responsibilities, discuss who their customer and suppliers are and how well needs are being met, and propose feedback process to ensure continuous improvement. The process als involves checking these perceptions by contacting the organization' customers and suppliers directly to obtain their inputs.

Performing a departmental task analysis at the start of a qualit improvement effort is invaluable because it creates a baseline o organizational performance against which business processes can b analysed and measurement systems can be established to provid data-based decision making. DTA concentrates on answering th following four questions:

- Who are you?
- What do you do?
- Who do you serve?
- Who supplies you?

DTA focuses on your customers and on your ability to meet thei requirements. It is only when these requirements are sought out an agreed that organizational responsibilities, effective measurement sys tems and requirements from suppliers can be fully determined. Figur 3.1 shows the relationships that become defined as a result of DTA.

It may also be useful as part of departmental task analysis to create first draft of an overall flow chart of the organization's main functions Defining these functions and their relationships, and gaining agree ment from the organization's members on their functions, is valuabl

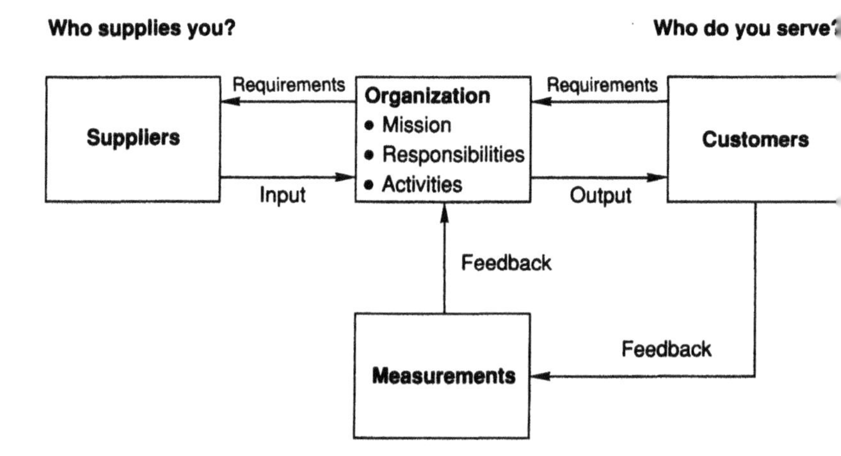

Figure 3.1 *Clarifying organizational relationships with DTA*

in clarifying what the organization does before writing the mission statement. The flow chart can then be refined as part of business process analysis.

Departmental task analysis is an eight-step process:

1 *Determine mission and responsibilities*: The first step is to develop a mission statement for the organization. This is often best accomplished where management and selected representatives (or all employees in the case of a small organization) sit down together to determine what they are in business to accomplish. It is especially important that everyone involved has a chance to have his or her say. If participants are not effectively made a part of the process, they may have little commitment to the resulting mission statement. Keep the statement relatively brief so that it can be understood and remembered by all members of the organization. Then it can be an effective tool to give direction to the organization, point to the primary 'customer' for its products and/or services, help members to understand priorities, and provide criteria for success. The mission statement should be written in general terms to answer the question, 'Why does the group exist?' Key words should focus on concepts such as improving, increasing or enhancing areas of the company's function. The statement should not be a description of what the organization or the individuals do but *what they are there to accomplish*. Questions that can be used to help define an organization's mission include:
   - What is our purpose?
   - Why does the company need us?
   - What business goals are we attempting to meet?
   - What business are we in?
   - What are we trying to achieve?

   Once the mission statement is complete, the next step is to define the organization's responsibilities. These are the natural 'what's' that flow from the mission statement. They are the tasks for which the group is accountable and is able to act on without guidance or approval from the next level of management. The mission statement should be reviewed and updated if necessary following step 4.
2 *Identify customers*: A customer is anyone inside or outside the company who uses products or services that are provided by an organization or by its individual members. People inside the organization who use its products or services are referred to as 'internal' customers. Anyone outside the organization is referred to as an 'external' customer. Primary customers are those people, groups, departments, etc. for whom the organization's products/services were primarily designed. Secondary customers, those who also use the products or services but who are not the primary focus of the organization, are also important. As part of the

customer-identification process all these primary and secondar customers should be listed. It may also be desirable to includ customer names and titles when these are available. In situation where an organization has a variety of functions and/or a numbe of separate missions, you may have different customers for diffe ent organizational responsibilities. Some questions you may cor sider in identifying your customers are the following:

- Whose needs are we trying to satisfy?
- Who determines the requirements for our mission and respons bilities?
- Whose interests are we primarily serving?
- Who is best able to evaluate our success?

3 *Determine customers' requirements*: To adequately meet your custc mer's requirements, you will need to meet with them on a regula basis to fully determine what their needs are and to track hov completely those needs are being met. Initially this should b done by having organization members identify customer require ments as they perceive them, but written in terms that customer would use. These are then taken to the customers and are checke with them. If multiple customers have the same requirements, it i useful to have them rank the relative importance of each. All cus tomer recommendations and requirements need to be compiled an prioritized based on justification and the number of customers wh have the requirement. Once this is accomplished, a mechanism should be established for continued communication with custo mers, so that further recommendations, success patterns and chang ing requirements can be adequately tracked.

4 *Determine activities necessary to meet customer requirements*: Whe you have determined and prioritized your customers' require ments, you and your group should define the specific activitie (either existing or proposed) needed to meet these requirements These may be processes or parts of larger processes that cut acros several organizations. Include tasks and flow times that direct th organization's resources to produce only what the customers need Before implementing these activities, revisit your primary cus tomers to review whether the planned activities will result in prod ucts and services that will satisfactorily meet their requirements. I is important at this stage to advise customers of any requirement that will not be met. Revise the mission statement if necessary tc ensure that customer requirements are fully met.

5 *Identify suppliers*: Internal suppliers consist of all other groups anc individuals within the company that provide products or service for the organization. Since many of these internal suppliers are no immediately obvious, your first task is to develop a full list of whc your suppliers are, both external and internal.

6 *Determine requirements to suppliers*: Once a full supplier list is determined, meet with the people in your organization to discuss your requirements for these suppliers and to collect information on how well these requirements are being met. Schedule meetings with your suppliers to communicate to them your requirements and recommendations. Create a communications channel within the organization to collect data on how well suppliers are meeting organizational requirements. In a parallel effort, develop a process to provide feedback to suppliers on how well they continue to meet these requirements.

7 *Design and implement feedback*: Developing an effective feedback mechanism is the key to a continued successful relationship with customers and suppliers. This mechanism should identify how information is to be obtained and provide a consistent, sustained flow of information sufficient to indicate quickly when the activities no longer meet customer requirements as measured by agreed-upon standards. Ask the following questions to determine the specific measurements to be used:
   - What will we look at to evaluate our own activities?
   - What will the customer look at to evaluate our activities?
   - What will we look at to evaluate our suppliers' activities?
   - What measurements will be used as indicators of success?
   - What data are available?
   - What additional feedback/data do we have to generate?
   - Do we know how well we are doing now?
   - What are our own targets for improvement?

8 *Identify defects*: Once customers' requirements and requirements to suppliers are fully articulated and feedback has been established, problem areas will be identified. Using the communication channels developed with both customers and suppliers, collect information on problem areas or potential problem areas and prioritize them for problem-solving efforts.

Once departmental task analysis has been accomplished, the organization is well positioned for business process analysis. This key step focuses on analysing the processes the organization performs to determine how they can be most effectively improved.

Exercise 3.2

Use DTA to define:

1 Your organization's activities
2 Your department's activities within the organizational context.

Business process analysis

Business process analysis is a method for analysing how work i accomplished to identify areas for improvement. Processes that ma be analysed range from hiring systems to drawing revision systems from software design to methods for procuring computers. Some pro cesses involve the entire company, while others are contained within small department or a single employee's job.

Any business process can be viewed as a group of related tasks tha utilize the resources of the business to produce specified results. A such, it is a repeatable sequence of activities that has measurabl input(s), value-added activities and measurable output(s) as show in Figure 3.2.

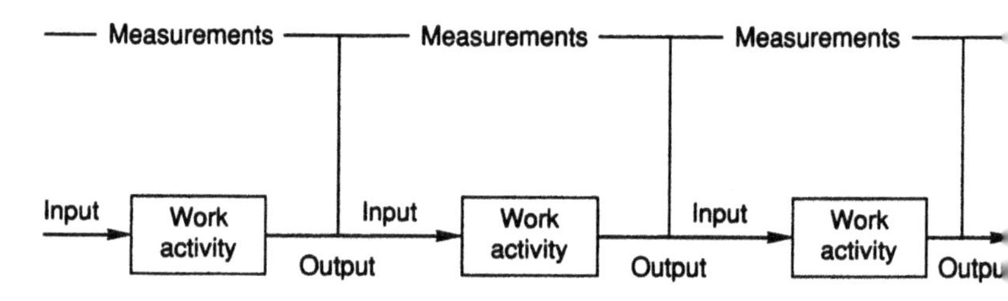

Figure 3.2 *Business processes*

The objectives of business process analysis are to find ways t improve effectiveness in meeting customer needs while eliminating waste of materials, capital and the time of people. This improves com petitive position while enhancing process adaptability to ensure con tinuing relevance to changing business needs.

A high-quality process has the following characteristics:

- Effective – achieves the intended results, meeting the customer's requirements
- Efficient – operates with minimum resources
- Under control – tasks are documented, responsibilities are clearly defined, variability is minimal
- Monitored – key control indicators are in use to identify changes in the process
- Value-added – contribution to business is defined, measured and tracked.

Achieving this level of process maturity is no accident. There are close similarities between managing processes and managing people, as is shown in Figure 3.3.

Both managing people and managing processes require a planned approach if maximum benefits are to be achieved. Too often, managers

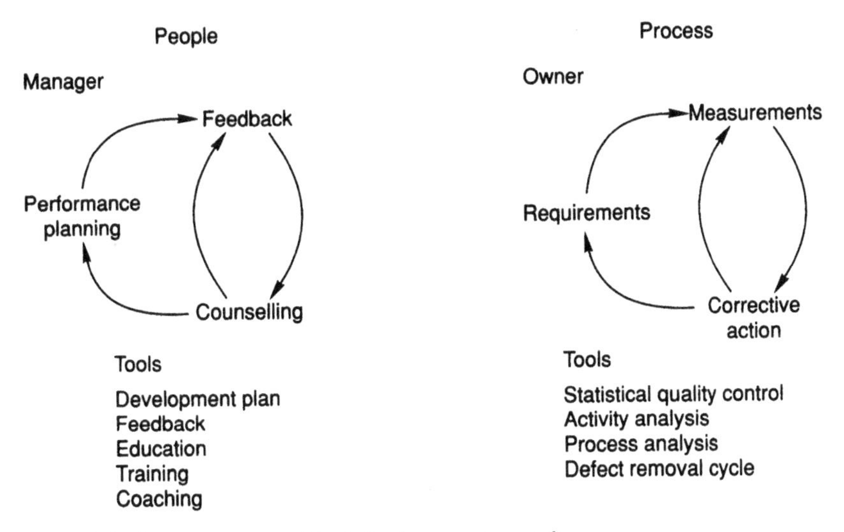

Figure 3.3 *Process management is similar to managing people*

try to improve a worker's performance or a process by jumping to a single tool as the answer.

Business process analysis can be implemented by the following seven steps. Each of these steps is important to ensure that real improvements are achieved and maintained for the long term. Taking short cuts in improving business processes is firefighting rather than real quality improvement, typically resulting in apparent short-term benefits that may actually increase costs in other areas of the organization:

1 *Name process and assign responsibility*: Determine which business process will be targeted for action. Candidates may be identified by listing the products of any organization. If a candidate includes several products that are dissimilar, look deeper to be sure that you have not described a function instead of a process. You should be able to describe where the process begins and ends. The process needs a name, consisting of a noun and a verb. Examples are 'promotion processing', 'software marketing', or 'graphics production'. At this stage a mission statement can also be written for the process if desired. In the early stages of quality improvement select a process that requires a minimum of involvement with other organizations and where the process flow is relatively straightforward. Early success provides encouragement to work the more difficult processes. Once a process has been identified, the next step is to assign process responsibility. The person given this responsibility should function at a high enough level in the organization so that they can:
   - Identify the impact of new requirements

- Commit to a plan and implement change for continuous process improvement
- Monitor the process efficiency and effectiveness.

When assigning process responsibility, other factors are also important, such as who in the organization has the greatest investment in the process; who does the most work in this area; who is affected by the outcome; who can best influence it; or who is measured by the quality of the process. The person who is given process responsibility ensures that all seven steps of business process analysis are implemented and that:

- Sub-processes are defined
- Line manager sub-process ownership is assigned and agreed upon
- Cross-functional issues are resolved. Once responsibility has been established, gain concurrence that he or she does in fact own the process and agree that it needs to be improved
- Finally, senior management above the process owner should inform other managers and organizations that operate within the process that improvement responsibility has been assigned and that the process has been targeted for improvement.

2 *Define process boundaries*: This should be done in writing and should include answers to the following three questions:

- Where does the process begin?
- What does it include?
- Where does it end?

In an order-processing system, for example, the manager who has been assigned process responsibility might define the process in the following steps:

- The process begins when a customer signs a letter to order equipment, software or services and the letter is date-stamped
- The process includes all steps, such as approvals and inventory checks, that are necessary to complete the paperwork in a timely manner
- The process ends when paperwork is completed and the work to be performed is fully specified

Next, the major suppliers and customers of the process need to be defined. If departmental task analysis has been performed, much of this has already been done. Work with customers to define the measurements that will be used to determine whether process outputs meet their requirements. The customers should also prioritize their requirements. Finally, implementers are defined. These are often people who manage a process; they are assigned the responsibility for determining process requirements and for identifying and removing defects from it. Because they manage tasks and resources operating within the process, they need to be intimately involved with its improvement. It is a strategic decision to select the process level where improvements will be made. The implementers might be sub-process owners; it might be more efficient to improve each sub-process individually, rather than work the process as a whole. However, if the improvement activities of implementers are highly interconnected, it is probably not wise to work at a lower level.

3 *Document process flow*: Once the process is named, responsibility assigned and the boundaries defined, the next step is to identify the scope of activity. Generally, this is defined as the business area that falls directly within management responsibility or the organization's charter. A high-level flow chart of the entire business process is then constructed. If possible, this should be constrained to a single page. The components of this flow chart consist only of major functional areas, decision points and external interfaces. Each major functional area and decision point is broken down into more detailed, secondary-level flow charts. The consensus of the organization's staff should be secured while creating these flow charts. It is surprising how differently people within a single organization can view the same process. This review alone can contribute significantly to the group's understanding of their individual and company responsibilities. Before becoming immersed in the details of each sub-process, develop a chart of the five to seven major elements that make up the process. This helps to focus the improvement effort in much the same way that making an outline helps in writing. In documenting process flow, the following stages are essential:
   - Specify the customers of the process and what tangible outputs the process delivers to them
   - Specify who the suppliers are for the process and what they provide for it to function correctly
   - Document the current process down to the task level. Process documentation includes all tasks within the process boundaries, presented in sequential order, along with the time required to accomplish them and who is responsible to make sure that they are accomplished
   - Use a process flow diagram to identify supplier and customer interfaces within the process. Define the 'added value' accomplished at each step in the process.

4 *Define control points/measurements*: Control points are places where adjustments in the process make significant differences in the quality of the process. Measurements are usually made further downstream in the process from the control points. This provides the data to determine what adjustment needs to be made at the control point. An accelerator pedal is an important control point in a car, for example, with the speedometer measuring the effects of its use. Between this control point and the measurement point is the engine,

a major process with a process. The brake, a control point farthe down in the car process, also controls speed. Predominant an smoother use of the accelerator as a control point and less frequer abrupt applications of the brake increases cost-effectiveness, a measured by petrol consumption. Business processes without mea surement systems can be like cars without speedometers. This ma lead to over-control, i.e. cycles of excessive acceleration followed b rapid braking and reduced cost-effectiveness. Control and measure ment points can be identified as shown in Figure 3.4. Each shoul be specified on the flow diagram. All measurements should b evaluated according to the question, 'What is done with th data?' Measurements need to become permanent parts of the sys tem when:

- They are used earlier in the flow to control the process and/or
- They are used to gauge the quality of the process, which is the used to identify areas for improvement.

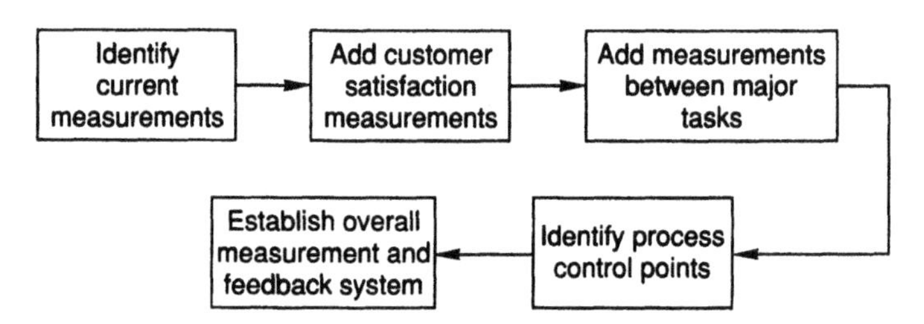

Figure 3.4 *Strategy to define control and measurement points*

Inputs from suppliers, outputs to customers and handoffs betweer internal organizations are logical places for measurements. Supplie measurements should be in terms of impact on the process Customer measurements need to demonstrate trends of satisfyin requirements. Measurements between internal organizations ar not made to fix blame but to give feedback early enough so tha corrective action can be taken.

5 *Communicate and implement*: Once the process has been definec and the control and measurement points have been identified, th process owner should communicate this with the people who wor in the process. Agreement is sought on who will make thes measurements and apply the controls. Informal processes, or work arounds, often exist because of failures within the main process These competing processes can then be disbanded, incorporatec into the regular process or allowed to exist until the formal proces is improved. If a workaround is allowed to continue, measure tc determine the costs of the workaround so that the way is preparec for its elimination. Most people would like to see the formal process

work; they just want to have solid evidence before they abandon their firefighting methods. Figure 3.5 shows the strategy required to communicate and implement the work accomplished to this point. Business processes must involve the people who work in the process. Otherwise, resistance may defeat the improvements. In fact, it is the people who work within the process who have the knowledge required to know where improvements really need to be made.

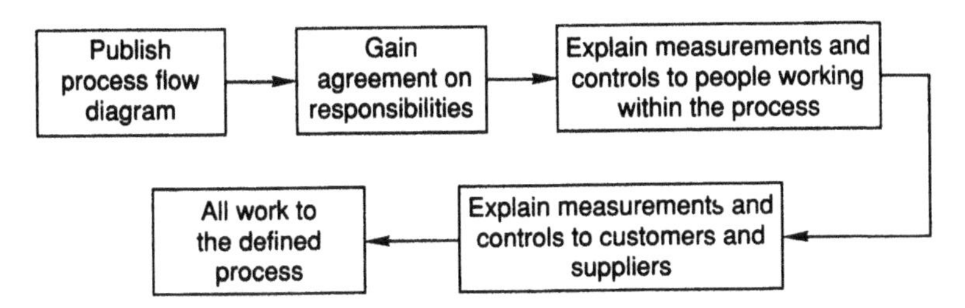

Figure 3.5 *Communicating and implementing the defined process*

6 *Measure and assess*: Very few processes can be analysed by using historical data that are already available. Most processes require new measurements, and new measurements require data before trends will become evident. This requires patience. If the prior steps have been performed well, improvements will be evident simply by defining the process and roles, but this is not enough. Continue to evaluate the measurement system to see if it is tracing customer requirements and revealing areas for improvement. You and your team, in conjunction with the people who use the process, should assess process performance on a regular basis. Your rating should be based on data that show how much the process has improved in operational effectiveness, efficiency and adaptability. A typical rating system is as follows:

Rating	*Score*
Process undefined, not measured	5
Process defined	4
Sustained improvement	3
Dramatic improvement	2
Defect free	1

7 *Identify defects*: Identify defects in the effectiveness and efficiency of the process as determined by the measurement feedback system. If a process is not effective or efficient, defects will be evidenced by:

- Customer requirements not being met
- Wrong supplier inputs
- Unnecessary tasks performed
- Redundant steps in the process.

Wherever firefighting and rework are found, defects are presen and where defects are present, repeatable patterns need to be dis covered. Sometimes this requires that the measurement system b changed to obtain sufficient data. It is often like a mystery that i solved by using the improvement process. Have the team prioritiz the problem areas and improvement opportunities that have bee identified. Assign an improvement team(s) to work with thes problems based on the priorities assigned. It is vital to communi cate that you intend to provide recognition for people who identif problem areas, rather than penalizing them for airing 'dirty laun dry'. A focus on determining guilty parties and assigning blame ca defeat the entire improvement process. The presence of defects ma indicate a need to redefine the organization's mission or roles and or the responsibilities of team members. If this occurs, it is useful t go back to the departmental task analysis phase to clarify this nev mission and the new roles and responsibilities.

Improving business processes

As defects are identified and their results are quantified, it is importan to take the time to determine true causes before pursuing correctiv actions. Once these true causes have been determined, changes in th process will need to occur. These are accomplished through problen solving.

The person who has process responsibility should review the statu of process improvement on a periodic basis so that it can be modified t accommodate new business needs and requirements as they occur Measurements are made to ensure that the process continues in th most effective and efficient manner.

Exercise 3.3

Use business process analysis to:

- Define a process you work with
- Determine measurements which will help to indicate the efficiency and effectiveness of the process.

Process re-engineering

There comes a point when the effort required to improve the perfor mance or efficiency of an existing process does not deliver an adequat return. Sometimes performance cannot be easily improved because o the impact or constraints of another department or function. Process re-engineering is an approach in business improvement that aims to change the way a business performs. It does this by focusing on the core or critical processes that deliver value to the customer.

To understand the current situation as regards efficiency of processes, it is important to identify value-added elements of the existing practices and to measure the performance of the business processes. Re-engineering a process means that the organization is realigned to deliver the value-added activities along the critical business processes, eliminating, as far as possible, the non-value-added activities.

Process re-engineering can be conducted at two levels:

1 *Process improvement*: Improvements are largely driven by internal factors and perspectives such as correcting recognized process problems
2 *Competing with competitors*: Improvements are largely driven by external factors and perspectives. Benchmarking is a technique used to determine performance gap. Process re-engineering can help to close the gap and make a quantum leap improvement.

Stages of process re-engineering

Stage 1: Determine key success factors and processes that produce them

The organization's strategy is often the starting point for process development as strategic planning can reveal the need for significant change and pinpoint the processes that need transformation. When choosing which process to redesign, the following questions should be considered:

- Could this process be eliminated (without affecting customer satisfaction)?
- Does the technology exist to replace this process?
- Is the process output required by an external customer?

Benchmarking can be an effective method of gaining a good understanding and knowledge of processes and practices. It is the act of defining the best systems, processes, procedures and practices. The measurement of business performance against the best of the best, through a continuous effort of constantly reviewing processes, practices and methods, can serve as an enabler for maintaining high levels of performance and competitiveness.

Types of benchmarking

There are, generally, four main types of benchmarking approaches which have been recognized:

1. *Internal benchmarking*: This is, strictly, a continuous effort of estab lishing good practice uniformly and company-wide by continu ously comparing what takes place in all the various operations c the organization. While the main advantage is the ease of imple mentation and the low requirement in terms of resources and tim it is targeting an internal standard only
2. *Competitive benchmarking*: This type aims at comparing specifi models or functions with main competitors. It has also bee described as reverse engineering (tear-down) since the startin point in most cases has been the product and looking at its char acteristics and functionality. The advantage of this approach is th direct comparison with main competitors. The disadvantage, how ever, is the difficulty with which information on processes i obtained and comparison with competitors may now point t best practice
3. *Functional benchmarking*: This type compares specific function with best in industry and best in class. It is a positive approach t benchmarking but, because it is related only to specific functions may not be of benefit to other operations in the business organiza tions concerned
4. *Generic benchmarking*: This is the ultimate in terms of benchmark ing application. This approach applies to all functions of busines operation. It encourages the continuous effort of comparing func tions and processes with those of best in class (see Figure 3.6).

Without benchmarking	With benchmarking
• Internally focused	• Understanding of other industries
• Few solutions	• Ideas from proven practices
• Customer requirements determined subjectively	• Best practice solutions
• Goals and objectives lack external focus	• Credible goals
• Strengths and weaknesses not understood	• Understanding outputs
	• Real problem solving

Figure 3.6 *Benchmarking*

Stage 2: Redesign of process

Once the gap between current performance and targeted performanc has been determined, the change process can begin. Naturally, a nar row gap can be resolved by a less radical change process, especially i the process is capable of being sufficiency improved without altering its fundamental structure. Where a wide gap exists, the process may have to be completely restructured before performance will improve.

Stage 3; Planning and implementation

Effective change management is extremely important during this phase. The aspect of change management which is critical is the assessment of the effects the redesign will have. A successful plan should include such elements as skill requirements, training needs, communication of rationale, and benefits of new processes to employees.

Problem solving

Identifying and solving problems is an integral part of all improvement projects and a means for continually satisfying the customer. The opportunities for improvement identified through departmental task analysis and business process analysis can provide possible projects for problem solving. Problem solving for quality improvement uses a variety of techniques that fall into four general categories:

- Flow charting
- Idea-generating techniques
- Problem-analysis techniques
- Statistical techniques (see Chapter 5).

The quality improvement tools relevant to the problem-solving process will be discussed in Chapter 7.

The problem-solving process (Figure 3.7)

1 *Identify improvement opportunities*: Estimate the customer and supplier quality requirements that are not currently being met. This can be accomplished by talking to customers and suppliers, asking people who provide products and services what their experiences are, etc. Another effective method is to survey the people in the organization to determine what they perceive to be their biggest problems and where the waste is (waste of people's time, materials or capital). Brainstorming is an effective tool for collecting a lot of information from a group in a short period of time. It is useful at this stage and at other stages of the problem-solving process. Identify the primary activities your department or unit performs that contribute to the business goals (see departmental task analysis). Which of these are reformatting or duplicating activities? Do they contribute unique elements to the end product or service? Using the flow charts created in business process analysis, determine the value-added steps and identify them on the flow charts. Value-added functions are any activities that directly contribute to the deliverable product. Not all non-value-added work can be eliminated, but it should be examined to compare what is really needed with what is being performed.

Step	Purpose	Process	Product
1 Identify improvement opportunities	Define a list of areas for improvement	• Employee surveys • DTA results • BPA results	List of potential improvement projects
2 Prioritize and select problems	Determine the most important problems for resolution	• Form a team to select projects • Choose according to greatest financial or quality impact	Prioritized list of projects
3 Define the problem	Determine the extent of the problem	• Use specific company language	Specific definition of the problem
4 Analyse problem causes	Determine the root causes of the problem	• State who, how, when, where, etc.	List of true causes
5 Develop solutions	Establish potential solutions	• Visualize the process without waste • Collect data • Determine true solutions	List of potential solutions
6 Prioritize and select the best solution	Determine the optimum solution	• Representation of all affected areas	Prioritized list of potential solutions
7 Implement solution	Solve the problem	• Develop an implementation plan • Represent all concerned • Maintain clear ownership of the problem	Best solution implemented
8 Evaluate improvement	Assess the degree of improvement	• Establish measurement system • Implement system • If not optimal solution repeat steps 3–7	Extent of improvement determined
9 Hold gains	Maintain the improvement	• Confirm ownership responsibility • Monitor process • Determine if it remains in control	Chronic area of waste eliminated

Figure 3.7 *The problem-solving process*

Examples of other items in the flow charts that can help identify areas for improvement are the following:

- Inspection
- Decision points
- Rework
- Reviews
- External interfaces
- Delays
- Customers
- Deliverable goods or services.

2 *Prioritize/select problems*: Prioritize the customer, supplier and employee quality requirements that are not being met. The problems that appear to be of highest common concern will often be the areas of greatest initial potential gain. Look for chronic areas of waste of people's time, capital and materials, and for work processes that can be improved. Select a problem to work on. Some solutions may be obvious at this point. If data exist showing that an optimum solution is already available, proceed to step 7. A key criterion in choosing areas to improve is the level of perceived financial or customer impact. These then become areas to investigate to determine whether the perceptions are accurate. Problems that span beyond the department or organization should be handled using a team approach involving membership from all affected organizations. Start by identifying a team leader and having them attend a team leader course, then form a team to work on the problem. The scope of the problem determines the membership of the team. Financial impacts can be determined by estimating the totalling the cost to perform the work, the cost of failures and the cost of the failure detection and measurement systems. This is referred to as doing a cost-of-quality analysis. If a significant problem exists, these numbers can be large and estimates may be all that is required to demonstrate that action needs to be taken.

3 *Define the problem*: Have the team members fully define the current problem. Use specific, understandable language, stating who, what, how much, when and where. State in terms of impact and answer the 'so what' question. Avoid terms like 'lack of' or 'inadequate'. The purpose is to make sure that all people concerned with solving the problem understand exactly what it is. To achieve this understanding, define the problem in writing and have the group discuss or modify the description before attempting to analyse causes. This ensures that everyone is working on the same problem. Have the team members fully document the current processes (use a process flow diagram). Initiate the process of gathering data on the problem by developing and implementing a data-collection plan. It is often advisable at this stage to consult with a statistician to make sure that the plan will result in useful and significant data.

4 *Analyse problem causes*: Before analysing causes, ensure representa tion from all departments and organizations impacted by the prob lem or its solution. Make sure the team members selected have th responsibility and authority to make process change decision Analysing problem causes is often best initiated by bringin together the team members and having them construct a cause and-effect diagram. This sets the stage for collecting sufficien data to enable true causes to be determined. Gather and reviev data to determine the amount of variation that is occurring withi the existing process. Use simple improvement tools as appropriate statistical tools as necessary. Data should be associated with way the customers and suppliers measure success of the product/se vice.

5 *Collect solutions*: Once the root causes have been determined an have been verified through data collection and analysis, the tean may choose to modify the process. Many modification ideas gen erally arise at this point. Solutions should be selected that improv conformance to customer needs and expectations, increase fitnes for use, and minimize variation from product to product, service t service.

6 *Prioritize and select the best solution*: A 'best' solution should b chosen for implementation, based on the data and a full investiga tion of the alternatives. Strive for solutions that involve group con sensus rather than having one or a few members of the tean dominate the discussion. When problems span beyond one depart ment, strive for solutions that will solve these problems across a affected organizations, then communicate this information upward

7 *Implement the 'best' solution*: Develop and implement a plan t eliminate the problem. This may involve modifying the process establishing training and education, job transfer, cross-training counselling, redefining company procedures, changing policies etc. When the improved process spans beyond the department communicate the implementation recommendations upwards i the organization. If higher-level approvals are required, manage ment presentations may need to be developed to acquire thes approvals. It is important to identify clearly who is responsibl for solving the problem, as well as who is going to follow through on implementation and evaluation.

8 *Evaluate improvement*: Measure to determine the amount o improvement that has taken place. Periodic measurement wi warn when a problem is returning or when gains are not bein held. The evaluation system adopted should be based on way the customer and supplier measure success. It should includ checks for conformance to specification, fitness for use and consis tency/variation. If real improvement was not achieved, return t steps 3 to 6 and repeat. Several repetitions of the improvement cycl may be required to achieve permanent improvement. The dat

collected initially may not accurately measure the amount of improvement for several reasons – fear of recrimination for 'failure', unclear data descriptions, wilful data misrepresentation due to pressure, desire to show immediate improvement, etc. Data are used in the new system:

- To reveal problems
- To verify the extent of a problem
- To analyse a problem
- To prevent problems
- To confirm that the corrective measures taken alleviate the problem or improve the process.

It is important to understand the points in a process where measurement and corrective action can produce the greatest benefits. These are called the key quality indicators. Discovering the key indicators in an organization or a process can be very challenging. However, it is one of the most rewarding activities a group can undertake in setting the stage for real improvement. A key quality indicator has the following characteristics:

- It is measurable, on either a numerical or a qualitative scale.
- It provides a basis on which decisions can be made that affect quality.
- It is understandable by those who need to review process improvement (including customers).
- It is expressed in terms that invite comparisons with similar processes.

Each process will have its own key quality indicators. Examples might be errors per page in a document, the flow time necessary to respond to customer inquiries, the rate at which employment applications can be processed, or the percentage of accounts payable that quality for a discount. It is important that indicators are selected at intermediate states within a process, where prevention-oriented quality improvements can be made, not just at the end of the process. In order to determine the amount of variation occurring in the key quality indicators of a process, measurement samples are taken according to a carefully devised measurement plan. This plan should always be constructed with the full participation of a person having substantial statistical training.

9 *Control at the new level*: Confirm responsibility for maintaining the improvement. Ensure that the improved process is continuously monitored and that data are gathered and reported (to indicate whether the process remains in control). Maintain visibility of improvements through normal business cycle reviews. This can be accomplished by establishing a system of periodic management review or quality audits of progress and performance. Perform the review in meetings already structured to obtain periodic visibility (such as during 'cost and schedules' meetings).

Conclusions

The process of satisfying and exceeding customer requirements may b depicted as in Figure 3.8.

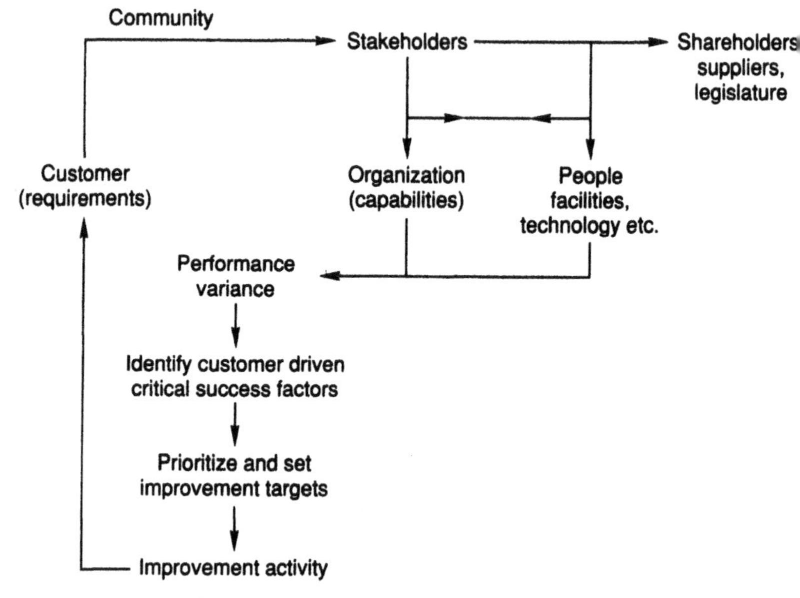

Figure 3.8 *Process of satisfying customer requirements*

This diagram suggest that there are many factors which impact ar organization, including shareholder direction setting, community feel ing, legislative requirements and supplier efficiency. These influencers of an organization may be termed stakeholders. Within the organiza tion itself, people skills and attitudes, facilities and technology al combine to deliver outputs of effectiveness and efficiency. Together the stakeholders and the organization define a capability or series o capabilities.

When compared to the customer requirements the organizationa capabilities may give rise to a performance variance. If this variance is positive (i.e. capability is such that it exceeds customer requirements then an organization may be satisfied with involvement in the continu ous improvement of its functions and processes. If, however, the var iance is negative (i.e. customer requirements exceed the organization's capability) then continuous improvement activity alone may not be enough.

It is therefore important to re-identify the customer-driven critical success factors. This may be performed by conducting BPA or DTA Having identified these critical success factors it is often necessary to prioritize and set targets. This prioritizing may be conducted using any number of common tools. Target setting is, however, more difficult, An improvement target of 20% reduction in cycle time may seem sound, but in an industry where 30% is the average reduction, this target will, in effect, move the organization backwards in relation to its competition. Benchmarking allows an organization to set targets based on best practice both inside and outside an industry sector.

Having set targets, continuous improvement activity may be sufficient to allow the organization to satisfy its customers. If the improvement required is quantum then process re-engineering may be essential. It should be stressed that in a total quality organization each step in the process depicted above is subject to a regular review of effectiveness and to continuous improvement in its own right. Thus there are many iterations and feedback loops.

ummary points

- It is essential to address the relationship between customer, supplier and organization.
- The management of these relationships using logical and structured tools and techniques enables an organization to understand its own capabilities and its customer requirements.
- The determination of customer requirements and an organization's own capabilities allows for realistic targets to be set.
- The change processes should be integrated into the strategic and operating planning processes.

4 Design control and management

Introduction

Design can be defined as the activity that converts the business plar and the ideas therein, into practical reality. It often requires the intel lectual and creative input of a number of people both within an outside the organization. The process begins with the translation c customer requirements into specifications and ends with handover t manufacturing or the service delivery function.

Design is fundamental to the creation of new products and service and is a key factor in ensuring a company's competitiveness and profit ability. Decisions taken during design influence product quality, cos and timing, and thus the market appeal, of the product. In addition, i often requires a high degree of innovation and consumes a significan amount of time and resources. There is therefore a need to manag design in a manner that affords the required degree of coordinatio and control without imposing unnecessary constraints on the creativ ability of the design team.

Establishing suitable control mechanisms which will operate effec tively over a long project can pose a problem for the manager. In thi chapter a methodology will be presented that will facilitate the contro and management of design in a manner that will ensure that th product offered meets the needs of the customer and the organization

Throughout the text the word 'product' will be used to describe both a product and service, and the objectives of the chapter are:

- To introduce the appropriate methodologies needed to manage an control design
- To discuss approaches to planning design activities
- To introduce the concepts of teamworking and concurrent engineer ing
- To discuss key tools and techniques commonly used in the desigr process.

The design process

Managing the design activity requires a set of principles common to al business activities. These underlying principles are based upon the need to organize, plan, monitor and control the various processes ir the organization. They encompass other management requirements such as setting objectives, communicating them to staff and motivating staff to achieve them. All such activities involve people, so their needs must be recognized and catered for in the management process.

In developing an approach to the management and control of design, the stages in the design process must be considered. This process often consists of four identifiable phases:

- Definition of objectives
- Conceptual design
- Project definition
- Detail design.

The clarity and importance of each phase will vary from company to company and from one project to another.

Definition of objectives phase

The first phase, sometimes known as 'definition of objectives', begins with the determination of the broad areas in which the organization wants to develop its products or services. This decision may be market-led, where product ideas are dictated by a stated customer requirement, or technology-led, where the organization may decide to develop an innovative product before market potential is evident. In practice, new products can result from either approach, or a combination of both, but should always emanate from an understanding of the market. Indeed, it is essential that customer requirements and other design input requirements are adequately understood before work begins. The organization can then develop a business case and decide on the viability of the idea before committing further spend on a project that may be doomed to failure.

During this phase overall business goals and objectives should be specified, along with details of requirements and constraints, including codes of practice and relevant legislation. All aspects of product development should be covered, including production, delivery and operation, as well as costs and timing. These design input requirements should be clearly defined in a formal working document, known as a design (or service) brief. This document should be made available to all those involved in the design activity and should be regularly reviewed and updated as the design develops. In addition, it should be used as a means for establishing a full planning, monitoring and evaluation process. Table 4.1 lists the contents of a typical design brief.

The importance of this phase cannot be overemphasized. Decisions taken on market needs and business objectives will have a major influence on the outcome of the project and yet the cost of these activities may be only a small percentage of the total cost of designing and delivering the product.

Table 4.1 The design brief

1	Aims and objectives
2	Company constraints (e.g. policies and procedures)
3	Special design considerations (e.g. legislation)
4	Marketing information
5	Flexibility requirements (e.g. interchangeability)
6	Functional requirements
7	Documentation (e.g. specifications)
8	Ergonomic requirements
9	Environmental requirements
10	Safety requirements
11	Production capabilities
12	Quantity requirements
13	Maintenance considerations
14	Packaging, transport and handling
15	Installation requirements
16	Physical dimensions and weight
17	Materials and treatments
18	Aesthetics and finish
19	Reliability issues
20	Standards and codes
21	Product-proving issues

Conceptual design phase

Once the overall objectives have been defined, a number of concept with the potential to fulfil them must be generated. It is vital tha concept options are generated to reduce the risk associated with th project and to ensure that only those concepts that meet the needs o the business and the customer are pursued further.

Competitors' products should be analysed during this phase anc relevant developments in new technology investigated. Concepts ca then be developed and carefully evaluated against the requirements o the design brief. The level of risk associated with each concept shoul also be assessed because, for example, some concepts may have goo market potential but may require substantial development funding, o may take a long time to develop. Once these issues have bee addressed, the most suitable concept can be selected for further devel opment.

Product definition phase

This phase lays the foundation for the final design phase through a structured development of the concept. The main functional aspect which will have a bearing on the way in which the product will per form should now be reflected in the concept. In addition, the method used to produce the product must be established and a detailed projec plan should be developed. The project plan should include a work breakdown structure identifying the various work packages, thos responsible for the work, the likely costs, timescales and review points. Full use should be made of techniques such as Network Analysis when developing the project plan.

The importance of this phase should not be underestimated. After a concept has been chosen there is often a tendency to rush into design before the concept has been further developed and the full implications studied. It is essential that the concept is reviewed to ensure that it will meet customer and business requirements, and that the level of risk inherent in the concept is acceptable, before detail design begins. Any changes that are required after this phase are likely to have severe cost implications.

Detail design phase

During the detail design phase the product will be fully defined in line with the design output requirements. All relevant documentation should be available, the design should be validated, and the production and delivery systems confirmed. Where applicable, full use should be made of computer-based simulation and analysis tools to verify that the design output complies with design input requirements. A final risk analysis should be completed before a decision to formally launch the product or service is taken. Sales decisions can also be made at this stage.

Managing the design project

The design process in any organization is likely to be well documented. However, problems can and do occur – communications can falter, coordination may slip, etc. Thus the management and control process must provide the organization with mechanisms for dealing with such problems in an effective and efficient manner, while enabling them to take advantage of unanticipated opportunities.

A suitable methodology based on a 'project management cycle' (PMC) is described below. The PMC is generic in nature and would be based on the organization's design phases, such as those described above. It consists of the establishment of a series of decision points in each phase at which management can review and control project progress (see Figure 4.1). For each phase a series of output requirements or 'deliverables' should be defined and these should be used as prerequisites to moving projects from phase to phase.

The PMC helps those working on the project understand what needs to be accomplished within each phase of the project and where they need to be by the end of the phase. At the same time, it provides management with the information required to determine if the project is ready to advance to the next phase. This, in turn, enables them to make a decision to release the resources required for the next phase.

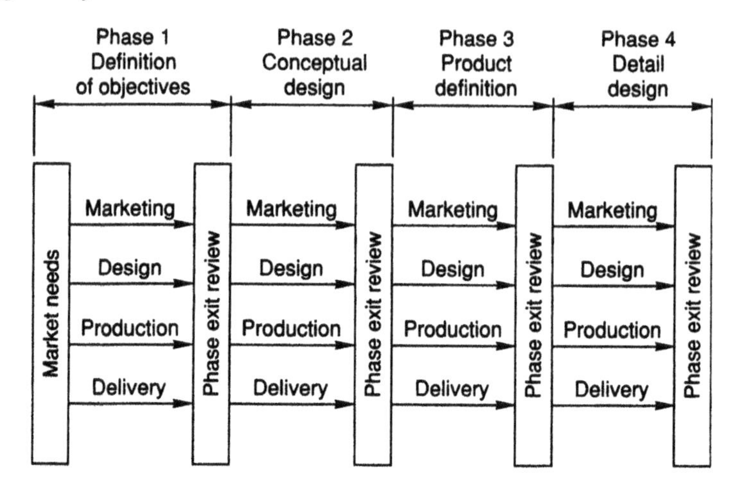

Figure 4.1 *The project management cycle*

Phase transfer decisions should be made during Phase Exit Review and should be based upon meeting the phase output requirement agreed prior to entering a particular phase. Formal design review should also be introduced to provide technical information for th phase exit reviews. These can be used to track the work accomplishe during a project of any size and to ensure that the needs of the busines and customer are met. They also enable the risk of entering the nex phase to be assessed before further resources are applied to the project

Phase exit reviews

Each phase exit review will assess project progress, data concerning th accomplishment of phase output requirements and plans for the nex phase. At these reviews, project readiness to enter the next PMC phas must be demonstrated not only in terms of the status of work but als considering the resource and skill requirements and facilities require for the next phase. The phase should not be exited until project readiness can be demonstrated. The objectives of the phase exit review include:

- Determination of the status of the project in relation to the plan
- Determination of the status of actions that have been agreed on, bu which deviate from the plan
- Confirmation of the resources expended during the phase
- Quantification of financial status against the plan
- Identification of current risk factors and their potential impact
- Identification of problems, their potential impact, and appraisal o plans for resolving them.

Phase exit reviews should be performed by a coordinating panel consisting of people with no daily involvement in completing the work during the phase but with the expertise necessary to review progress. This group would normally consist of functional managers and should be given responsibility for reviewing progress and making recommendations to proceed. Where ownership of phase activities changes, the chairperson of the coordinating panel may also change. The choice would depend upon the phase that is about to be entered and the nature of the work, to ensure that the requirements for entering the phase have been met before proceeding.

Issues may arise during these reviews which cannot be resolved by the team and which must be passed to senior management. The important point is that these issues are identified and that action is taken to resolve them before proceeding to the next stage of the design activity. Unfortunately, in many organizations problems are allowed to accumulate before action is taken. This leads to increased cost and risk as the project progresses.

The person responsible for coordinating the work of the design teams should be involved in the reviews. This will ensure that he or she takes ownership of the project and that communication remains effective between the various parties.

An important part of the preparation for the phase exit review should include a self-assessment by the design team. They will know more about the project than anyone else, so it would be beneficial to allow them to assess project readiness for the formal review. The output of the self-assessment would be a proposal to the coordinating panel containing proof of conformance to the output requirements. Where the output requirements are incomplete the team would be expected to produce a recovery plan to complete the work for the exit review. This would ensure that resources are applied to areas of need so that problems can be rectified in a timely manner.

Before authorizing phase transfer the assessment groups must satisfy themselves that the technical and business requirements have been met in full. If they consider that they have not been met then the conditions to enable the project to enter the next phase should be identified. The assessment panels must also agree the phase output requirements for the next phase with the project team coordinator.

Design reviews

Design reviews can be used as an objective evaluation procedure to ensure that a design will meet all its technical requirements. They help to minimize risk by providing a means of identifying inadequacies in the design, so that corrective actions can be taken to ensure that the design meets the requirements identified in the design brief. They also help to avoid repeating problems that have been made on previous projects and to indicate the likely product cost and profit potential.

In the context of the PMC, design reviews could be used to satisfy the organization and its customers of project progress, and as an additional input into the phase exit reviews. The reviews should be conducted by a team comprising representatives of the relevant disciplines, the personnel designing the product, sub-contractor and customer representatives. This will ensure that an objective evaluation of the product is achieved prior to the phase exit review.

Design reviews should be scheduled during the design phases and should address product, production and delivery process issues. The number of reviews would depend upon the amount of work to be completed during the phases, and the time and risk involved. The purpose of the concept design review would be critically to assess the concept options in terms of their ability to satisfy the requirements of the design brief. Potential problems should be reviewed and a formal technical risk assessment of each concept completed.

The purpose of the project definition design review is critically to assess the compliance of the selected concept with the design brief and to reassess areas of technical risk. The specifications to be imposed on sub-contractors should be examined at this review to ensure that they are complete and agreed.

The design phase design review should critically review the design to determine if it meets the requirements of the design brief. This review will concentrate on a critical appraisal of the performance of the product during tests or proving trials. The viability of the project should be considered and a further risk assessment conducted. In addition, the status of the design package and associated documentation should be reviewed and the availability of all information required for production and delivery should be confirmed.

A procedure should be prepared for both phase exit reviews and design reviews. An agenda should be prepared for each review and the results of the meetings, including action plans, should be recorded. This information and the resulting actions taken should be formally reviewed during subsequent phase exit and design reviews.

Exercise 4.1

Identify the main stages in your organization's design process and suitable review points. Develop a series of deliverables for each review point.

Project planning

Where design projects involve a number of people or teams a project plan must be developed to facilitate the management and control of their activities. This should include a plan of activities, timescales and resource requirements. Depending upon the complexity of the project, it may be necessary to employ established planning techniques, such as Gantt charts or Network Analysis, when developing the plan. Where the project is relatively simple a bar or Gantt chart can be used to create a plan together with an associated schedule that can then be used to monitor progress. The Gantt chart shows the proposed start and finish of each activity graphically against a common timescale and helps to identify priority areas.

For more complex projects Network Analysis can be used. This method is particularly useful where the project consists of many interdependent activities as it shows the interrelationship of the various tasks which make up the overall project. It also identifies the critical paths through the project, while providing information on time, cost and resource aspects of the project.

The object of the technique is to represent the activities of the project in their logical sequence, connected together using nodes to form a network. This in itself is useful as it forces the user to identify the particular activities involved in the project and any sequencing problems. Time analysis can be used to determine start and finish times for each activity and key milestones can be identified. The longest path through the network can then be found. This is important, as this path will control the completion date of the project.

The real power of this method is that, once the logic of the project activities has been defined, a microcomputer can be used not only to calculate and analyse times but also to assist the user with resource scheduling, calculation of spend profiles and barchart representations for review purposes. As with all such methods, the use of the computer will allow revisions of the network to be completed quickly, which is particularly important for design projects where changes can often occur.

Within the context of the management and control process discussed above, the techniques described can be used by the project coordinator and team leader when planning the project and by the design and review teams to monitor progress. Network Analysis can be used to define resource requirements for each phase and to agree on deliverables and review dates for each phase. Such techniques also act as useful methods for communication between all those involved in the design project.

Concurrent engineering

The sequencing and timing of the various activities required to produce the phase output requirements is vital. This has resulted in the emergence of a a range of approaches to product design. The traditional sequential approach used by many organizations involves each activity from design through production to delivery being completed one step at a time, in virtual isolation from the others. The advantage of this approach is that it is easily managed, although it is time consuming since each downstream group has to wait for the preceding one to complete its tasks. In addition, it ties up scarce resources, reduces the amount of feedback from downstream functions and results in a lack of accountability for the results of the process.

The suitability of this approach in today's competitive environment is highly questionable. There is now a need for an approach that facilitates the design of the product, production and delivery processes to take place in parallel if the phase output requirements are to be met in a timely and cost effective fashion. This approach, known as concurrent engineering, is based on the development of close working relationships between all functions. It ensures that all elements related to the provision of a product are considered from the outset and that decisions are based on consideration of downstream as well as upstream inputs.

The success of concurrent engineering depends upon teamwork and effective communication within and between the various groups involved in the project. In addition, it is essential that each person involved is committed to meeting the project goals. One of the most effective approaches is based upon multi-disciplinary teams coordinated by a project leader. This structure is discussed below.

Multi-disciplinary teams

The single most effective element of concurrent engineering is teamwork. The use of multi-disciplinary teams will enable product, process and support system design to be conducted in parallel. The ability to incorporate the combined knowledge of the team members in the design will ensure that problems are solved at source, minimizing the level of change and improving quality while reducing costs and cycle time.

The teams should be given the responsibility for planning the activities required to meet the phase output requirements (to a programme agreed between themselves and the project leader) and executing the plan. This programme should be sufficiently stretching to encourage the team members to adopt a concurrent rather than a serial approach to task completion. Giving the teams the authority and responsibility for completing the required tasks combined with pre-planning will ensure that they take ownership of the project. They should also have the freedom to define the work steps, inputs and outputs required to meet the phase output requirements. The success of this approach should be recorded and reviewed at a later date to determine those actions which proved successful and those areas requiring further attention.

It is important that the teams do not involve themselves in activities that are required for a later phase because, in essence, they are wasting time. The phased approach should ensure that this does not happen and that the teams concentrate on 'doing the right things'.

eam structure

Any team management structure must ensure that teams can work effectively and that they get the management support they need. A popular organizational form consists of a series of teams, the number depending on the complexity of the project. A team leader should be appointed to coordinate the contributions of the team members and to ensure that those tasks relevant to the team and the phase of the PMC are completed within agreed timescales. The team leader must be given the necessary authority to set priorities and allocate resources so that control over the project is maintained within the team.

A project leader should be appointed with complete responsibility for the project, for coordinating the activities of the teams on a day-to-day basis and for achieving, through the teams, the phase output requirements and meeting the programme. The project leader should compile the project plan and assign dates for the project phases in conjunction with the team leaders. They should define the resources expended during each phase and the resource and skill requirements for the next one. The project leader should also assemble the teams for each phase through agreement with the functional heads and departmental managers. To ensure that the project leader has the required level of authority he or she should report to senior management.

The project leader and team leaders should, where possible, be full-time and should remain with the project throughout each phase of the PMC. This will ensure that loyalty remains with the one project and that the transition from phase to phase is smooth, i.e. the learning-curve effect of handovers is minimized. It will also enable the team leaders to gain extensive project experience which will be useful when they eventually move on to another project.

The teams should consist of individuals from the relevant functional departments working in peer groups, sharing responsibility and ownership for the success of the project. The fact that the teams are multi-disciplinary will ensure that adequate attention is given to all aspects of the project during design (e.g. delivery issues are considered alongside product design). The ability to address such issues at the working level is one of the main advantages of the concurrent engineering approach.

The functions represented on the teams will depend upon the nature of the work in each phase and the phase output requirements. Team membership should change from phase to phase to ensure that the

relevant knowledge is available within the team. Bringing new idea
and knowledge into the team is important and such information shoul
be disseminated throughout the team. Team membership should be o
a full-time basis if even for only a short period during the phase.

The roles and responsibilities of the team members should be docu
mented, although this should not restrict their ability to contribute t
each other's work. This is essential if the product and related processe
are to be developed concurrently and if tasks are to be completed o
time. In addition, the increased flexibility will give the team member
the opportunity to assume more and wider responsibilities, which wi
improve their career prospects and motivation within the team.

For teams to be really effective they should be co-located from th
start of the project. This will aid inter- and cross-team coordination
improve the speed of decision making and ensure that the team mem
bers communicate efficiently and effectively. It will help to break dow
any barriers that exist between the departments represented on th
teams and build the trust that is necessary if the team approach an
concurrent engineering is to be successful. In addition, it will enabl
each team member to gain an understanding of each other's point o
view so that decisions taken reflect the varied needs of the team mem
bers.

Another benefit of the co-location approach is that the team leade
will have the necessary control over the team members to ensure task
are completed concurrently and on time. Concurrent engineering is les
successful if the team acts as a committee where work is assigned to a
team member who takes it back to his or her functional department
This approach simply adds another layer of management and furthe
complicates communications because of the need for numerous pro
gress meetings.

By co-locating the team, they will eventually break away from ol
patterns and develop new ones which are more conducive to bringing
products to market both efficiently and effectively. Co-locatio
encourages team members to work in parallel and to identify oppor
tunities for overlapping. Information can be supplied when required
enabling the team members to start their tasks sooner. The search fo
ways to get started earlier encourages a problem-solving approach and
keeps the focus on overall project goals. It also encourages the sharing
of responsibility and ensures that the design cycle is compressed.

However, where customer and supplier representatives are members
of the team it may be undesirable to co-locate them. In addition those
team members requiring access to equipment not available in the team
area and specialists may not be co-located. The advantage of keeping
specialist staff together is that a specialist knowledge base can be
maintained and that new staff can receive adequate training and
experience in the working environment. The flexibility for working
on various tasks can also be provided this way. Where this situation

exists, the team leader should continue to coordinate the work of the specialist. The people concerned should also be encouraged to spend as much time with the team as possible so that they feel part of the core team.

Although the team members will report to the team leader during their involvement with a project the functional manager will continue to retain title to them and will still be responsible for their professional development. This approach will give the team leader the authority needed to reduce development times while ensuring that the team members are secure in the knowledge that they have a home to go to when their involvement in the project ends. This is essential if the team members are to remain committed to the team and its goals rather than solely to their functional department. It will also encourage functional management to support the proposals.

Information technology

Information technology (IT) is a term used to describe a range of technologies with applications in data analysis and information handling. These technologies can be applied to any part of the organization that produces or uses data and so have wide application across the organization.

Design may be considered an iterative process, with each iteration generating an increasing level of information. Coordinating the collection, processing, storage and dissemination of information is essential for effective design. In the concurrent engineering environment, timely access to accurate information is essential. The use of teams and a suitable coordinating structure, as described above, will help to ensure that communication and information exchange is effective. However, additional steps may be required to further reduce the likelihood of problems as a result of poor communication. The use of information technology offers many advantages in this environment and many organizations are now exploiting the possibilities offered by IT systems, such as computer-aided design (CAD), to ease information flows.

CAD can be considered as any activity that involves the effective use of the computer to create or modify a design. One advantage of CAD is that it provides a means of storing and sharing information efficiently and effectively while reducing errors and resultant waste and rework. Past designs can be retrieved for modification or new design work helping to reduce cycle times while improving response times for answering design queries. Other advantages include:

- The ability to analyse design options quickly and accurately enable designers to generate a greater number of potential design solutions This provides the user with the opportunity to explore innovativ ideas and therefore acts as a stimulus to creativity
- The ability to integrate CAD with other computerized simulatio and analysis techniques means that products can be modelled on computer reducing the need for a large number of physical model and testing. This facility also frees the designer from the need t perform tedious calculations during the design process, releasin him or her for more creative work
- The ability to check the effect of modifications prior to implementa tion and to respond quickly to customer queries
- Reduced costs and higher-quality products
- Increased customer confidence in the capabilities of the organizatio

Perhaps the greatest advantage of CAD is that it can improve th degree of integration between departments and teams which is essen tial for the successful introduction of concurrent engineering practices

Quality function deployment

The need to gain a thorough understanding of customer requirements before design begins has already been discussed. If these requirement are not understood then no amount of design control and managemen will ensure that the customer receives a product that meets his or he needs. Quality function deployment (QFD) is a structured method ology used to capture the needs of the customer and to translat them into appropriate business requirements. QFD was first applie as a design management technique on a US Navy contract a Mitsubishi's Kobe Shipyards in 1972. It was introduced into the USA in 1983, and Europe in 1988, and has broad application in manufactur ing industry, public and service sector services. Current users includ Ford, Toyota, Rank Xerox, Procter & Gamble and Mars.

QFD was originally created for new product development althoug it can also be used when improving existing products. It can be applie during each stage of the design, production and delivery process t highlight areas of concern to the customer which must be addressed i the resulting product is to prove successful in the marketplace.

The methodology is based on a four matrices which are used t convert external customer requirements into specific design and manu facturing quality characteristics. The first is the product plannin matrix, often called the 'House of Quality' (see Figure 4.2), and is com pleted by following an eight-step process. The first step is to identify potential customers and their relative importance. Competing product should also be identified at this stage and all the information should b

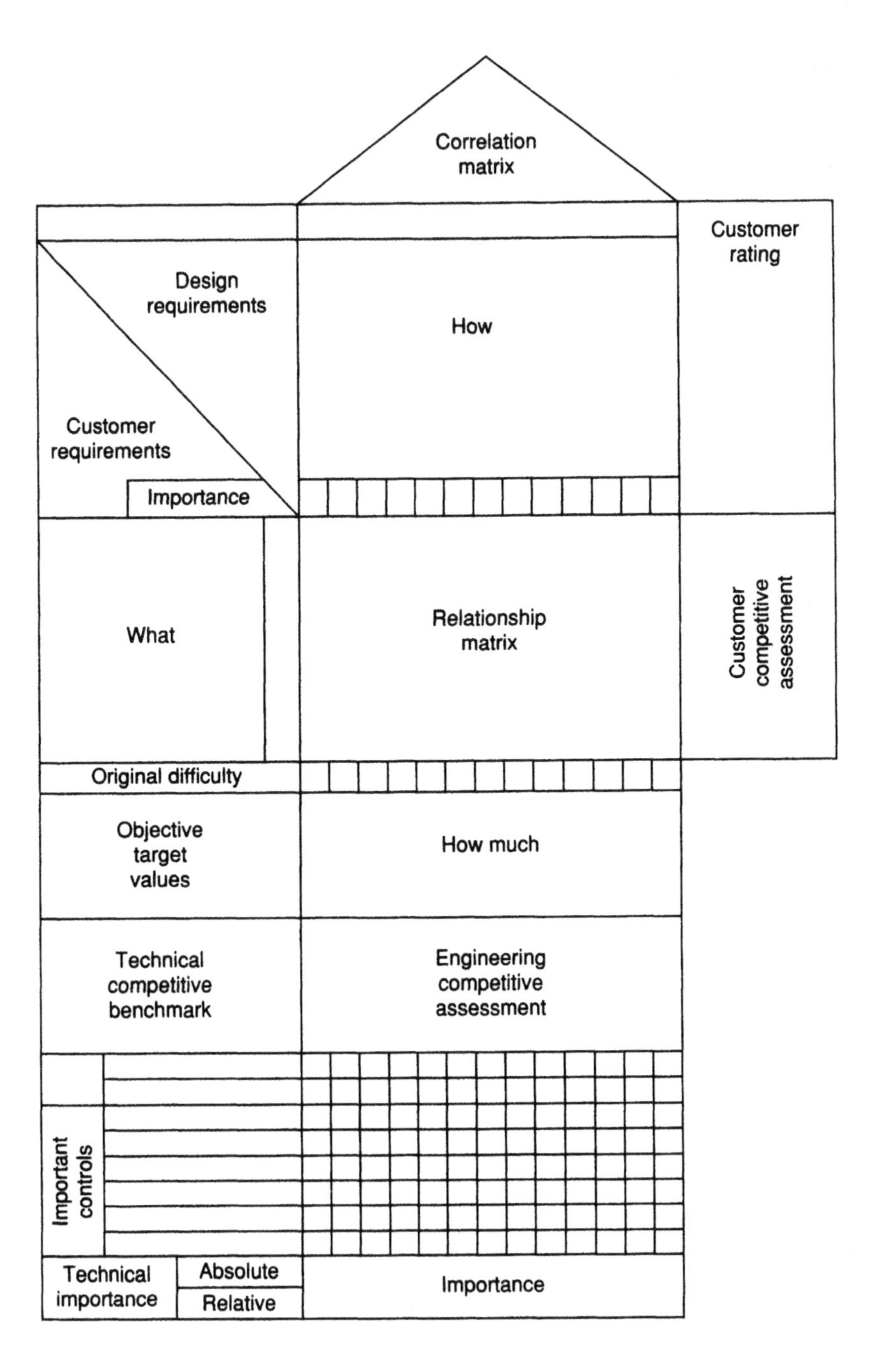

Figure 4.2 *House of quality*

recorded in the matrix. In Step 2 customer requirements, stated in their terms, are recorded in the matrix. This information can be collected through discussion with customers, competitor analysis and market intelligence. This is the most critical part of the process and it usually is the most difficult because it relies on the user obtaining and expressing what the customer truly wants – not what the organization thinks he or she expects. This information is then used to drive the design process.

In the third step customer-expressed importance ratings for the listed requirements and competitive evaluation data for existing products are recorded. The competitive evaluation data indicate how the customer rates the organization's existing product, compared to competitors' product offerings, and those areas requiring improvement. The average importance and improvement factors are then computed and sales points are identified. Sales points are characteristics which could be advertised and are to be emphasized in a particular market segment. These data are used to calculate an overall importance rating and the percentage contribution of each customer requirement.

Step 4 in developing the matrix is to list ways in which these requirements may be met that will predict customer satisfaction. These design requirements establish those features which should drive each stage of the product development process.

In Step 5 the relationships between the customer requirements and design requirements are defined. The significance of each relationship is then quantified and the matrix is analysed to determine whether the design requirements adequately cover the customer requirements. If, for example, there is an absence of relationships, or there are a high proportion of weak ones, this indicates that some customer requirements have not been addressed or that the design requirements, and therefore the design, are unlikely to satisfy those particular customer needs. Conflicting design requirements can also be identified using this aspect of the relationship matrix. Once the relationships are known, the design requirements may have to be modified or supplemented to ensure that the customer requirements are adequately addressed.

In Step 6 the importance of each design requirement is calculated to prioritize them. The relationships between the various requirements can then be studied and trade-offs identified and resolved.

In Step 7, target values for the design requirements are defined. These are based on the selling points, the customer-importance index and an evaluation of the competitive position of the organization's and competing products. The data should be expressed in measurable terms before, as the final step in the process, they are compared to the market evaluations to determine areas of inconsistency between the customer's evaluation and that of the organization. Where inconsistencies exist these should be investigated to determine if the internal evaluation is flawed or if the wrong characteristic was chosen to meet a particular need.

In the next stages of QFD the final product characteristics are deployed through a Part Deployment matrix which identifies those finished component characteristics that must be controlled to ensure that customer requirements are met (see Figure 4.3). Customer requirements can then be further deployed through a Process Planning matrix to examine the process used to produce the components to identify suitable control and monitoring points within the process. These then become the basis for operating instructions and the process control strategy defined in a Production Planning matrix. Customer requirements are therefore fully deployed so ensuring that those points critical to meeting them are controlled throughout the design, production and delivery processes.

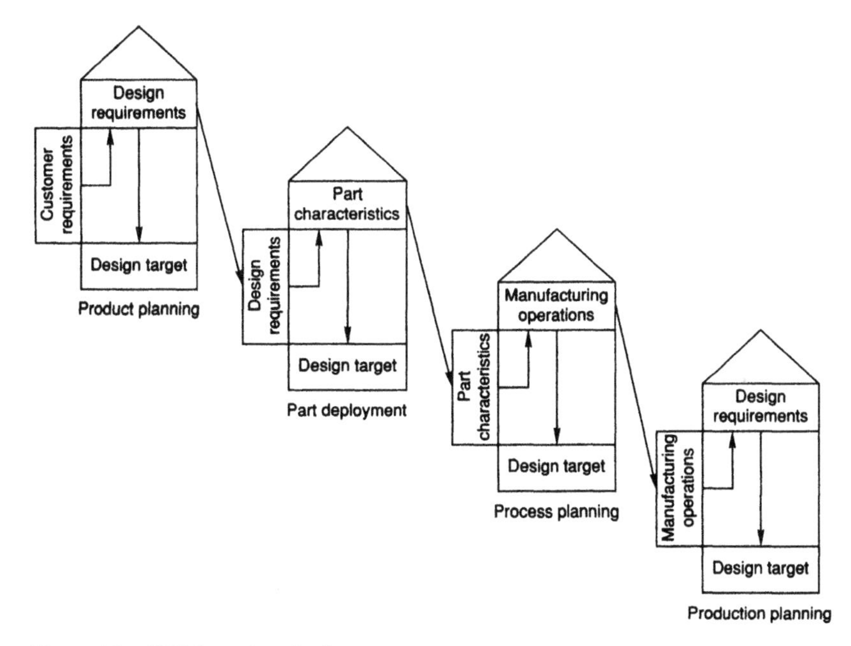

Figure 4.3 *QFD in product development*

The success of this approach will depend on the time invested in the early planning stages. The additional time required to study customer requirements, and the relationships between the design requirements, will be offset by reductions in overall development time and product changes. In addition, the competitive assessment aspect of QFD can be used to compare competitive products and to target selected areas where competitive advantage can be achieved. As such, the technique would be useful in the early phases of the PMC. Furthermore, as QFD encourages information sharing and increases the efficiency of information transfer by creating a disciplined outline for all involved to follow, its use would ensure that information is not lost during each phase of the PMC.

The nature of the methodology requires and indeed promotes a tean approach by fostering a better and common understanding of require ments and what is needed to meet them. This can contribute to reduc tions in change levels, so contributing to reductions in lead time. All c these issues are vital to the success of any concurrent engineering effor

Other advantages of QFD include:

- Fewer product-launch problems
- Lower product-launch costs
- Knowledge transfer.

Additionally, QFD provides a tracking system that identifies th impact of changes to a product. It also establishes a database for futur design improvements and improves communications between al members of the team. As a result, it is one of many techniques tha can be used for quality improvement.

Conclusions

A methodology for the management and control of design, known a the 'project management cycle', has been presented. It consists of a number of phases which establish decision points at which manage ment can review and control project progress. The method can be use to track the work accomplished during a project of any size and t ensure that the needs of the business and of the customer are met. I also enables the technical and financial risk of entering the next phas to be assessed before further resources are applied to the project. I addition, it facilitates the evaluation and continuous improvement o the entire design process.

For each phase a series of agreed output requirements must be me before the project can move to the next phase. It is also recommende that additional effort should be placed in the early phases of the PMC This is essential because, although adopting a concurrent engineerin approach will reduce the time to market, it is pointless developing th wrong product for the market. This approach will ensure that th customer and business needs are identified before concurrent engineer ing is employed to produce a product in reduced time.

Suitable project planning techniques have been presented, particu larly Network Analysis, which may prove useful when plannin design tasks. This technique enables the user to identify the tim needed to complete design activities. The importance of time ir today's competitive marketplace calls for new approaches to organiz ing for design and a new methodology, known as concurrent engineer ing, has been presented. This approach relies heavily on teamwork anc a suitable team-based structure has been discussed.

Where possible, customers and suppliers should be encouraged to join the design team to ensure that their requirements are reflected in the design. A technique, known as quality function deployment which translates customer requirements into terms that are meaningful to the organization, has also been presented. This is a powerful methodology that can contribute to design management and control, especially when combined with the use of a design brief.

The above structures will make a positive contribution to the management and control of design in the organization. They will also improve teamwork and communications and stimulate creativity, which is essential to business success. When combined with information technology that supports such activities, the organization will be capable of designing products in an effective and efficient manner.

Summary points

- The project management cycle can be used to manage and control both product and service design projects.
- Project progress should be monitored by phase exit and, where appropriate, design reviews.
- The sequential approach to design should be replaced by concurrent engineering supported by multi-disciplinary teams.
- A range of tools and techniques such as computer-aided design and quality function deployment can be used within the design process.

5 Control of quality

Introduction

If one of the key reasons for improving quality is to achieve increase customer satisfaction then increased customer satisfaction must b measurable, and this requires that quality itself must be measurabl In order to maintain a competitive price and ensure a reliable produc which performs to specification and which also has aesthetic appeal, i is imperative that the process involved in the design and manufactur of the product, and the inputs to these processes, are understood an are in control. Processes must be stable, predictable and capable.

In an ideal world materials and information would flow into a organization and would be used directly where they were neede (e.g. raw materials directly onto the shopfloor or custom-designe software directly to the point of use. However, one of the man malaises of Western industry and commerce is that we build an accep table quality level (AQL) into our systems. In reality, we are saying tha we expect and can accept a certain percentage of defective product. It i not untypical to have an AQL of 95%. Thus, in this instance we ar saying that we are happy with 5% of our product being wrong firs time. The objectives of this chapter are:

- To review techniques of controlling variability
- To illustrate that defect prevention is the goal of company-wid quality management programmes
- To explain that statistical process control is a major prevention tech nique by which a process is monitored and controlled
- To illustrate that process improvement is an integral part of th change process.

Inspection

There are many systems which can be put into place to minimize th amount of scrap and rework (wrong first time) output from a system Perhaps the simplest is incoming goods inspection. This is put in plac for two key reasons. First, there is an understanding that by lettin poor-quality goods enter our system it is difficult to recover the situa tion and produce a quality product. Second, we are saying that we d not trust our suppliers and thus our own vendor-rating system Perhaps the most significant outcome of this scenario is that ou organization must perform non-value-added activity in order to achieve an acceptable quality of incoming goods. Through the employment of goods inwards inspectors a very high price is being paid for a product which should have been delivered correctly in the first place.

Another way of describing non-value-added activity is cost incurring. Thus inspection, in general, is a cost-incurring exercise. In an attempt to work in a more cost-effective manner sampling inspection is preferred to 100% inspection, and in many situations 100% inspection would be impossible.

As inspection is non-value added it is important to understand why an organization should perform it at all. There are very positive reasons (e.g. because a customer demands it) so that decisions can be made about what to do with the materials next or to identify and thus correct a defective process. There are also certain fundamentals which apply to any inspection process, i.e:

- How items are measured
- How inspection errors are classified
- How inspection is costed
- The methodology for inspection
- Inspection diagnostics.

Measurement may be applied to two different forms of data: attribute and variables. Inspection as applied to attributes essentially means sorting product into a small number of nominal categories (e.g. pass/fail, good/bad, ripe/unripe, on/off, etc.). Inspection using variables data is concerned with absolute information (e.g. the weight, length, temperature, concentration, etc.). These measurements can be used to make the overall accept/reject decision. No matter which inspection system is used, it should be cost related and must be an integral part of the overall manufacturing process and understood by all employees who affect the process.

Sampling

Sampling is used where the cost of inspection is high or the proportion of defectives is low. Based on the number of defects found in a sample, a decision can be made as to whether the entire batch should be accepted or rejected.

Sampling has many advantages and some disadvantages. Compared with 100% inspection advantages include:

1. Economy
2. Less handling damage
3. Lower labour costs
4. Lot rejection as opposed to defectives rejection.

There are, however, some disadvantages:

1 Risk associated with accepting bad lots and rejecting good lots
2 Additional documentation
3 Normally less information about the product.

In developing sampling plans two key assumptions are made: firs that inspection teams follow the sampling plan and second, that inspec tion is made without human or machine error. In practice, thes assumptions are not valid. The results of sampling are greatly influ enced by the way in which a sample is selected. In acceptance sam pling, the sample must be representative of the lot. Published samplin tables have been designed on the assumption that samples are draw at random, i.e. at any time each of the remaining uninspected units c product has an equal chance of being the next unit selected for th sample. Random sampling requires that random numbers be gene ated and that these random numbers be applied to the product to b inspected.

Unless rigorous procedures are designed for random sampling, th sampling can regress to a variety of biases which are detrimental to th decision-making process. Common biases include:

1 Sampling from the same location in all containers
2 Previewing the product and selecting only those units which ar defective or non-defective
3 Ignoring those portions of the lot which are difficult to sample.

Many sampling plans set up their criteria for assessing lot conformanc in terms of the allowable number of defects in a sample. Defects ca vary greatly in the degree of seriousness and sampling plans shoul take this into account. Thus a separate sampling plan is used for eac classification of defect severity.

Standard quality control texts discuss operating characteristic curve and the most common forms of sampling plan in depth. However, th application of any acceptance sampling plan to a particular lot require information on three key elements:

1 The sample size (i.e. the number of items to be sampled from a lot
2 A statistic (i.e. a value derived from the sample which serves as a index of lot quality)
3 Decision criteria (i.e. the specification of values of the statistic whic lead to acceptance or rejection).

In implementing the sampling plan a sample of the stated size is take and the statistic computed from the sample results and compared wit the decision criteria. This offers the simple choice of accept/reject.

ailure mode and effects analysis (FMEA)

FMEA was developed in the 1920s and was not effectively used until the 1960s, when it became widespread in the aerospace industry. It has been used as a technique for preventing defects and for improving quality. In simple terms, FMEA provides a subjective assessment of potential causes of failure and also an analysis of the effects that such potential causes could have during the product lifecycle. A basic seven-step process exists for conducting FMEA:

Step 1 involves the identification of a specific part or process and subsequently identifying its function as accurately as possible.
Step 2 requires the identification of the potential failure mode or deviation. The deviation needs to be described by answering a series of questions, namely:

1 What could go wrong with the part of process?
2 In what way could the part fail to conform with specification?
3 Even if specification is achieved, what could the customer find unacceptable about the part?

Typical failure modes could include broken parts, poor surface finish, distortion, leaking, etc.

Step 3 is based on the assumption that the deviation has occurred. The effect of the deviation needs to be described from a customer perspective. Effects could include, excess noise, lack of power, not function, etc.

Step 4 looks for the causes of the effects described in step 3. A list of potential causes is assigned to each deviation (e.g. wrong material, worn tools, assembly error, machine set-up).

Step 5 deals with the ratings which need to be applied to the system. The rating is a subjective assessment of the degree to which each of the items highlighted above can affect the part of the process. Ratings are agreed by team consensus and again, as in many other areas of quality management, group dynamics becomes critical to the process. Typically, the severity of the effect of failure on the customer is rated on a 1–10 scale:

- The complexity is rated on a 1-5 scale. This is based on the number of parts or processes involved in producing the final item being analysed
- An estimate of the number of occurrences of the particular failure mode is rated on a 1-10 scale.

Finally, an estimate of the probability that a defect caused by the particular failure mode will reach the customer is rated from 1 to 5. This is essentially an estimate of the integrity of the quality system.

Step 6 continues the previous measures of severity, complexity occur rence and detection into a factor known as the criticality index (or risl priority number, RPN). The criticality index can thus be used to asses the design of a part or the process associated with manufacturing th part.

Step 7 requires corrective action to be taken on high RPN values Subsequent to corrective action, a new RPN value should be produced

The value of this technique is to ensure that all risks are identifie and action taken. Thus FMEA should give a high degree of confidenc that a product will satisfy customers and, equally importantly, b designed for manufacture, thus reducing internal conflict.

Statistical process control (SPC)

The management philosophy of W.E. Deming[1] is based upon a focu on variability and the causes of variation in a process, product o service which prevent its performance from remaining constant in statistical sense. Deming has estimated that 85% of quality problem are due to faults in the system rather than faults of the employee. Thus blaming workers is de-emphasized, and responsibility for product o service quality is placed largely in the hands of management Management's task is to provide or develop the organizational systen (e.g. authorize the purchase of better tools or raw materials) tha enables workers to perform to maximum capacity. Developing th system also includes creating a culture where workers feel secur enough to ask questions, report trouble, or make suggestions fo improving the product. Deming insists that managers think of th process as extending beyond the manufacturing facilities to includ suppliers, dealers, engineering marketing and management. The entir system can be partitioned into many interrelated sub-processes, an SPC can be applied at any stage in the process, or to the entire process

The most fundamental principle of SPC is that all processes produc variation in the output. Deming teaches that variation which occur both in tangible products and in services has two sources: normal an special. Normal variation reflects numerous natural, extraneous an unsystematic factors (normal causes) that are inherent in the system Such factors can be regarded as 'random' or 'noise' in the system Normal variation can be reduced but it can never be eliminated Because it is random and inherent to the process, it is often difficul for management to accept as the cause of poor quality. A partial list o factors in the system that can cause normal variation in the outpu includes:

1 A poorly trained group of employees
2 Unreliable instruments

3 Vibration
4 Humidity
5 Poor product design
6 Slight inconsistencies in the raw materials.

Special cause variation, in contrast, reflects the existence of special or unusual, non-random factors affecting the system's performance. These special factors affect the system such that it produces outcomes that differ greatly from the modal outcome. Special variation generally arises on an irregular basis and influences only some of the product. A broken tool or an inappropriate adjustment are examples of factors that produce special variation. Often, although not always, special cause variation is caused by factors which can be corrected at the machine by the operator. If the variation is non-random, some intervention is called for to bring the process back into statistical control. This is done by identifying the special causes and eliminating them.

A primary objective of SPC is to determine whether variation in a product or service is due to normal or special causes. If the variation is caused entirely by normal causes, then the system is 'in statistical control'. When a system is 'in control' the variation in its output will assume a normal distribution, and it will be stable over time (see Figure 5.1).

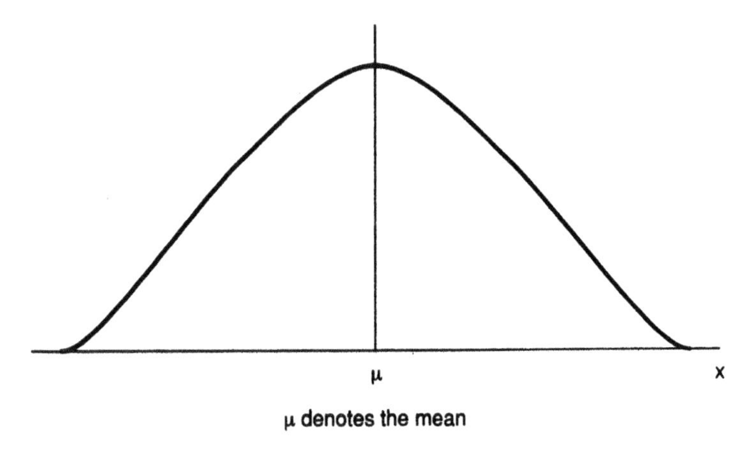

Figure 5.1 *Normal distribution*

Statistical control is not a natural condition for a process. It must be achieved by eliminating the special causes in the process. When all the special causes have been eliminated, only the random, but predictable, variation remains.

An unstable process cannot be improved. Attempts to improve a chaotic system will be confounded by the special causes already in the system, making it difficult to know the impact of the intervention.

Therefore management's initial task is always to bring the process int statistical control. Once a process is in statistical control, managemen can systematically work on shifting the process average closer to th desired level or reducing the amount of normal variation in the syster Management must realize that there is no quick fix to system problems Reducing the variation caused by inadequate training, poor produc design and inappropriate methods may require years to show improve ment.

When a process is in statistical control it does not imply that th output is acceptable, only that the output is consistent and predictabl (i.e. the percentage of non-conforming parts can be determined befor they are produced). The variability may still be too large, and have t be reduced before acceptable products can be consistently produced However, since the process is stable, the problem(s) must lie within th system itself and cannot be solved by simple actions by the operator Only management can solve such problems.

After statistical control has been achieved, special causes may occa sionally re-enter the system and cause it to go out of statistical control This will be evident by changes in the process average (Figure 5.2), in the process variability (Figure 5.3), or in both the means and variance (Figure 5.4).

Failure to distinguish between normal and special cause variation has two unfortunate consequences. First, when normal variation i mistaken for special consequences, attempts by the operator t improve quality by greater effort or by tampering with the curren work procedures are prone to failure. This only results in increase variability, higher costs and increased frustration as workers and man agers try unsuccessfully to 'remedy' the system. Second, when norma variation is attributed to the worker rather than to the factors inheren in the system, the worker can be incorrectly rewarded or punished fo results that are beyond his or her control. Failure to consider th natural variability that is inherent in the system may lead the manage to make inaccurate reasonings about the causes of performance and, in

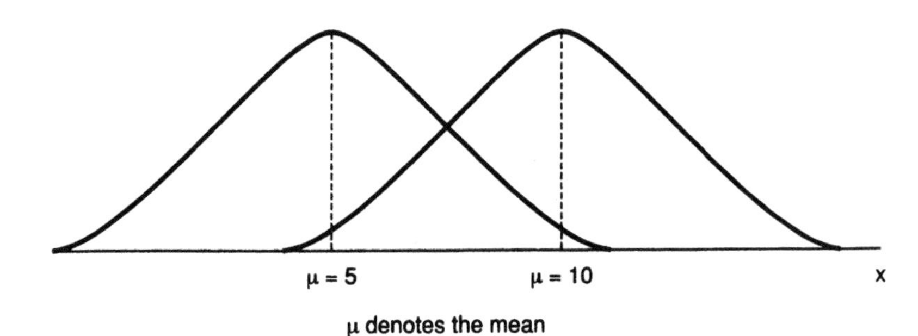

Figure 5.2 *Changes in the process average (different means)*

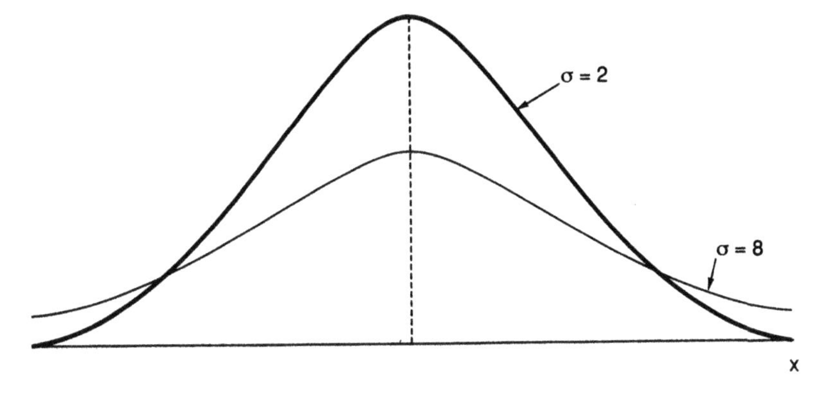

Figure 5.3 *Changes in the process variability (different standard deviations)*

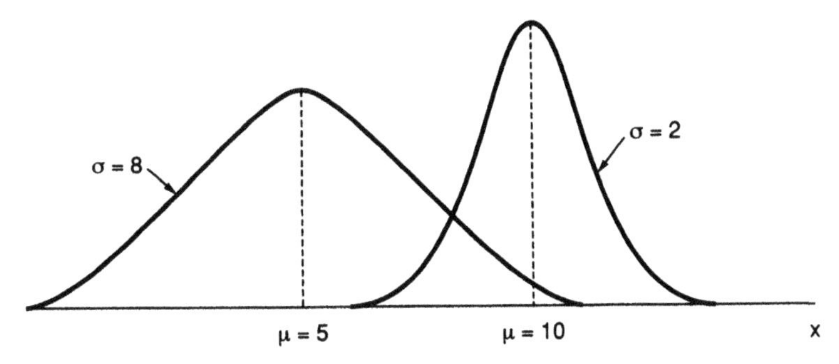

Figure 5.4 *Changes in both the means and variances (different means and different standard deviations)*

turn, to reinforce behaviour in ways that are both counterproductive and unfair.

Reinforcement schemes that simply reinforce natural random changes in behaviour will be ineffective, and cause managers to become cynical of performance-contingent reward systems. At the same time, subordinates are likely to feel confused, competitive and unfairly rewarded. Deming has been a long-time ardent critic of traditional performance appraisals and merit systems.

All employees whose performance puts them in the 'within' category should receive pay rises based on seniority. No one in this category should be rewarded or punished based on their performance. To do so would reduce morale and cooperation and cause the variability of the system to increase as below-average workers overadjusted their behaviour. Employees who perform on the 'high side' should be given merit pay and their methods should be taught to other employees. Employees who perform on the 'low side' should receive coaching, Individuals who consistently perform on the 'low side' should be replaced.

This approach assumes that almost everyone is within the system. management is dissatisfied with the average level of performance c spread of the distribution, then it must remove performance barrie and change the system.

Control charts

The primary tool for determining whether product variability is norm or due to special causes is the control chart. There are two types c control charts:

1 For attributes (used where the dimension either does or does nc conform to standards)
2 For variables (used where the dimension is measured as a contir uous variable such as length or weight).

Both types of charts serve as signalling devices that tell the worke when to intervene and remove special causes, and, just as importan when to leave the process alone. Without objective data and this char ing method, it is easy to confuse normal and special cause variatior While several variations of these control charts exist, the essentia features of all charts can be illustrated by discussing the *X* bar and control chart.

The *X* bar and *R* control chart simultaneously tracks both the mea (*X*) bar and the dispersion (range of *R*) of a process, and displays then on graphs for easy interpretation. Each important dimension of th product (critical to the customer or where frequent defects are found would require a separate *X* bar and *R* control chart. Therefore a proces may be monitored with several control charts – one for each importan quality dimension. The *X* bar and *R* control chart consists of tw graphs: an average chart and a range chart. To construct the *X* ba and *R* control chart, same-sized samples are drawn at regular interval from the process or output. The product dimension (diameter, weigh length, etc.) of each piece in each sample is measured. The sample mea (and range) of each sample is calculated and plotted, in temporal order on special graph paper. Next, the grand mean (and average range) i calculated, and drawn as the centre line on the *X* bar chart (*R* contro chart). A typical *X* bar and *R* control chart is shown in Figure 5.5.

To determine whether or not a process is in statistical control, on must determine:

1 Whether the sample means are normally distributed around th grand mean; or
2 Whether the sample means deviate significantly from the gran mean.

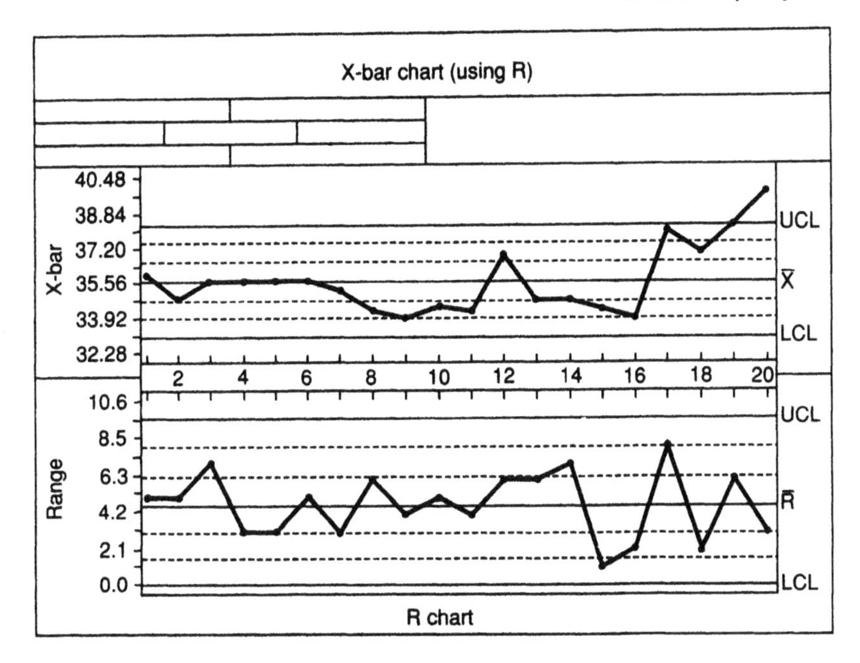

Figure 5.5 *X-bar chart and R chart*

If either the operating level or the dispersion vary by more than could be expected by chance alone, the process is said to be out of statistical control. Significant deviations are indicated by values that lie outside upper and lower control limits.

Typically, the upper control limit (UCL) is set at three standard deviations above the grand mean; the lower control limit (LCL) is set at three standard deviations below the grand mean. By chance alone, only about three out of 1000 outcomes would lie outside these control limits. Thus any sample value lying outside these control limits suggests the presence of special cause variation.

To aid the analysis of control charts, they may be broken into three distinct zones:

- Zone C is ±1 standard deviation either side of the grand mean
- Zone B is between ±1 standard deviation and ±2 standard deviations either side of the grand mean
- Zone A is ±3 standard deviations either side of the grand mean (see Figure 5.6).

If the process is in control the sample means would be normally distributed about the grand mean. Therefore, approximately 68% of the sample means would fall into the two Zone C's (each one being one standard deviation from the mean), approximately 95% would fall within Zones B and C, and approximately 99.7% would fall within Zones A, B and C. Furthermore, the sample means should

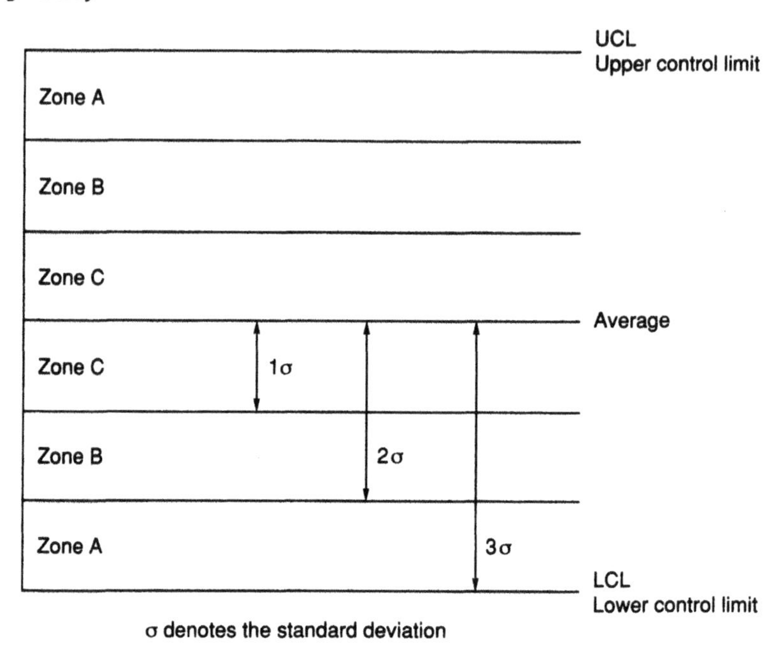

Figure 5.6 *A, B and C zones for a control chart*

vary randomly and not exhibit any trends or cycles. There are a serie of common rules applying to control charts which, if violated, tell th user that the process is out of control. The eight most common rule are:

Rule 1 One point beyond Zone A
Rule 2 Eight points in a row on the same side of the centre line
Rule 3 Eight points in a row steadily increasing or decreasing
Rule 4 Fourteen points in a row alternating up and down
Rule 5 Two out of three points in a row in Zone A or beyond
Rule 6 Four out of five points in a row in Zone B or beyond
Rule 7 Fifteen points in a row in Zone C, above and below the centr line
Rule 8 Eight points in a row on both sides of the centre line with non in Zone C.

If any of the tests described above are violated, then the process i probably being affected by a special cause. For example, in Rule 7 while a control chart may indicate random variation about the mean the process is out of control because more than 68% of the sampl means fall within Zone C. There are several interpretations of such chart:

1 There has been a significant improvement in the process resulting ir less variation;

2 The control limits were calculated incorrectly (perhaps because data from two processes are being inappropriately mixed together); or
3 Operators may be editing the data because they have previously been punished when the process was out of control.

Several points about SPC are illustrated by this example. First, the control chart indicates the presence of a special cause, but it does not necessarily identify what it is. Second, interpretation requires that both the X bar and R control charts be examined simultaneously. If data from two different processes are being mixed the sample means will 'hug' the grand mean on the X bar chart, but the R control chart will show considerable variability. Finally, managers and operators must perceive control charts as problem-solving aids and not as devices for policing operators.

Companies without SPC try to ensure quality parts by mass inspecting the finished product and discarding or reworking the defective ones, This is wasteful and unreliable, and it does nothing to improve the process. SPC, in contrast, attempts to implement a defect-prevention strategy which emphasizes 'doing it right the first time' and eliminating the need for costly and unreliable inspection of the final product. In order to 'do it right the first time' the process must be:

1 In statistical control
2 'Capable'.

The capability of a process is determined by comparing actual process performance against specified engineering specification for the item being produced. Ideally, the actual variation in the process should be less than 75% of the product tolerance, and centred between the specification limits. This will allow the process to produce acceptable parts even if it should occasionally go out of control (e.g. a shift in the process mean).

Figures 5.7–5.9 show three processes with different degrees of capability. In Figure 5.7 the process is in control but not capable of producing all parts to specification. Over time, this process will consistently produce, on average, 45 defective units for every 1000 units produced. The operator is powerless to reduce this number. Only management can make the changes that will reduce the variability in the process. Comparing Figure 5.7 with Figure 5.9 shows that improving the capability of the process (i.e. reducing the spread of the distribution) will decrease the number of non-conforming products. Moreover, if the process does go out of statistical control (i.e. a shift in either the process mean or the process variability), the process will still produce good parts (assuming the change is not too drastic). If the process is being monitored with a control chart, then shifts in the mean and variability can be detected and corrective action taken before the process begins to produce bad parts. Thus, once a process has been made capable, the

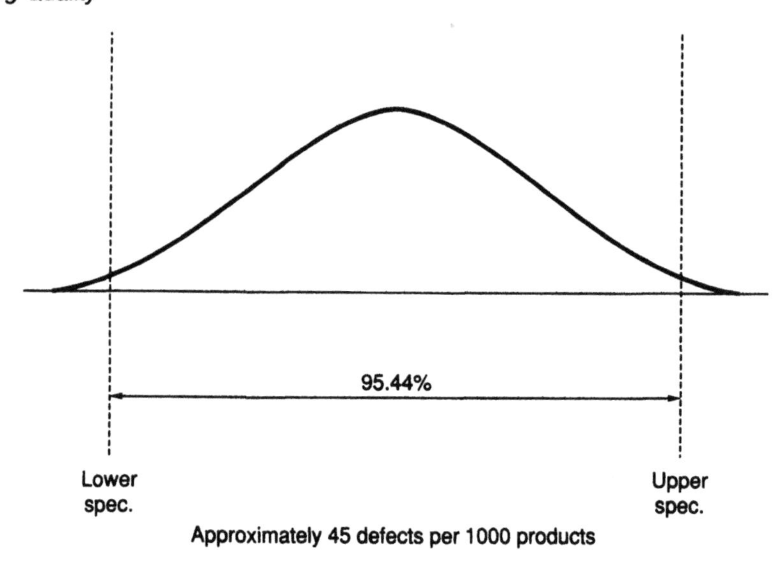

Figure 5.7 *Incapable process*

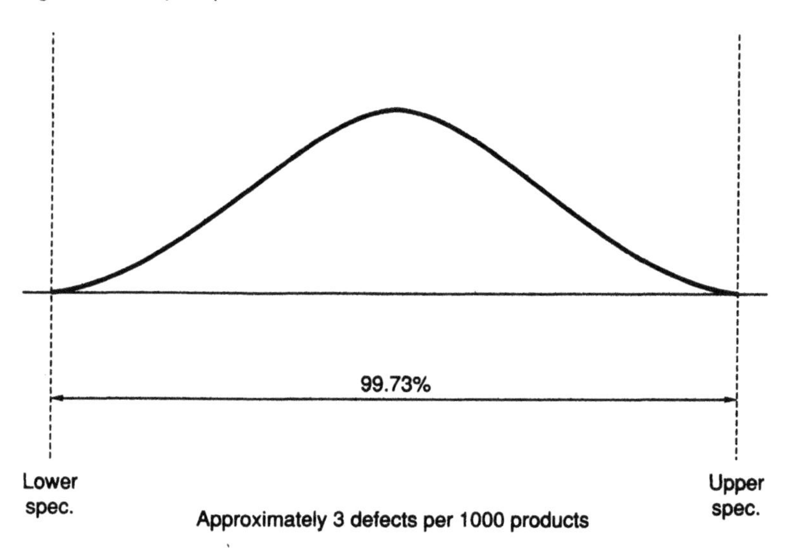

Figure 5.8 *Marginally capable process*

control chart acts as an early-warning device that identifies when a special cause has entered the system. If detected early enough, the special cause may be found and removed before very many (if any) non-conforming products are produced. This is known as defect prevention through process control.

The control chart simply tracks the process, and identifies what is happening with the process mean and range. It identifies when a special cause has entered the system, but it does not necessarily identify

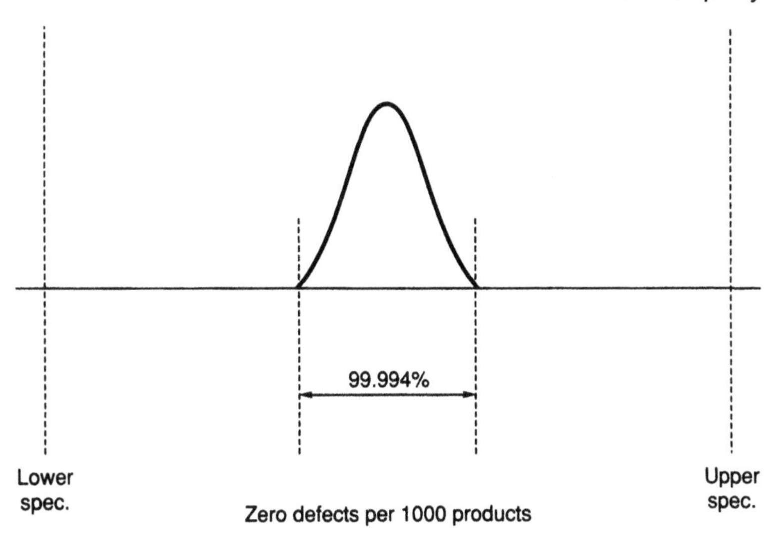

Figure 5.9 *Highly capable process*

what the special cause is. However, sometimes the special cause can be easily determined because the control chart displays a known pattern or the special cause comes and goes at regular intervals. In such cases, the operator may routinely intervene to remove or prevent the special cause.

For example, by tracking processes with control charts an operator may periodically change tools before the process goes out of control due to tool wear and thus prevent the production of defects. Similarly, other patterns on a control chart would suggest other causes (e.g. operator fatigue, defective raw material, changes in machine settings, etc.).

Management must go beyond using control charts to signal when in-process adjustments are needed. Even with extensive charting nothing very significant or very long-term will happen if management does not practise employee involvement and teamwork and push for never-ending improvement (i.e. constant reduction of process variability). Control charts, then, are simply the first step of an ongoing problem-solving process.

Worked example

By way of a worked example, imagine a situation where you work for a fast-food delivery service. You are responsible for tracking the time it takes to meet the delivery schedule. Table 5.1 shows the time taken in minutes to reach the customer across 100 deliveries. Five consecutive deliveries are shown as one sample.

Table 5.1

Sample no.	*Value 1*	*Value 2*	*Value 3*	*Value 4*	*Value 5*	*X bar*	*Range*
1	33.00	36.00	37.00	38.00	36.00		
2	35.00	35.00	32.00	37.00	35.00		
3	37.00	35.00	36.00	36.00	34.00		
4	37.00	35.00	36.00	36.00	34.00		
5	34.00	35.00	36.00	36.00	34.00		
6	34.00	33.00	38.00	35.00	38.00		
7	34.00	36.00	37.00	35.00	34.00		
8	36.00	37.00	35.00	32.00	31.00		
9	34.00	34.00	32.00	34.00	36.00		
10	34.00	35.00	37.00	34.00	32.00		
11	34.00	34.00	35.00	36.00	32.00		
12	35.00	35.00	41.00	38.00	35.00		
13	36.00	36.00	37.00	31.00	34.00		
14	35.00	35.00	34.00	34.00	34.00		
15	35.00	35.00	34.00	34.00	34.00		
16	33.00	33.00	35.00	35.00	34.00		
17	34.00	40.00	36.00	38.00	42.00		
18	38.00	36.00	37.00	37.00	37.00		
19	35.00	39.00	41.00	38.00	38.00		
20	39.00	40.00	42.00	39.00	39.00		

For a sample of size 5: $A_2 = 0.577$
$D_3 = 0.000$
$D_4 = 2.114$
$d_2 = 2.236$

For the *X* bar chart the mean is 35.65
the UCL is 38.27
and the LCL is 33.03

For the range chart the mean is 4.55 the UCL is 9.62
and the LCL is 0.00

By looking at the *X* bar and *R* control chart (Figure 5.10) it is easy to identify the out-of-control situations. It should be stated that the *R* control chart is in control, thus we can move on to look at the *X* bar chart. As an example from point 2 onwards we violate Rule 2. Points 17–19 violate Rule 5 and point 20 violates Rule 1.

Conclusions

Statistical process control may be viewed as a strategic change mechanism or a vehicle for strategic change in that it:

- Focuses on the continuous improvement of every business process which affects customer satisfaction, whether that be in a manufacturing environment or in the service side of a business. This must, by necessity, be a long-term objective
- Requires a change in management philosophy away from productivity and towards quality. This also requires a change in management style

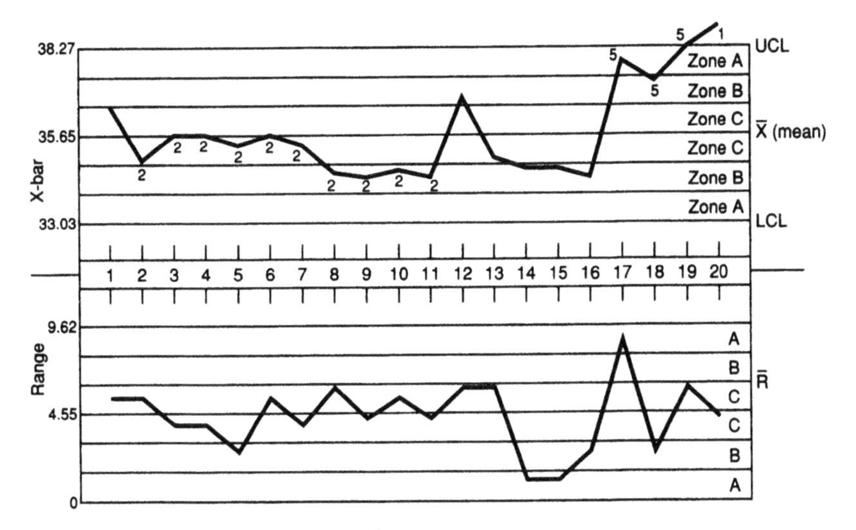

Figure 5.10 *X-bar and R chart*

- Seeks to gain competitive advantage through higher quality, better cost base, higher return on utilized assets, etc.
- Requires commitment from all levels within the organization and sound project management
- Addresses employee involvement and teamwork
- Requires extensive training and retraining of operators and managers alike
- Challenges the fabric of the organization by requiring an evaluation and improvement of traditional employee-appraisal systems.

Summary points

- Statistical process control is a method of monitoring and controlling a process instead of inspecting products.
- Variation in processes and products can be measured.
- There are two types of data: attribute and variable.
- There are two types of course: chance and assignable.
- Process capability is a study to determine whether specification limits can be attained by a process or machine.
- Capable processes can be improved through management action.

Reference

1 Deming, Walter, E. (1986), *Out of the Crisis*. Cambridge, Mass.: MIT, Centre for Advanced Engineering Study.

6 Managing quality costs

Introduction

One natural outcome of the recognition of the importance of qualit has been an increased focus on improving quality through systems standards and continuous improvement. The pursuit of these activitie requires considerable effort and incurs cost. However, the cost asso ciated with errors and the subsequent need to rectify them are usuall an order of magnitude greater. In recognition of the existence and magnitude of these costs, many organizations have begun to measur and act upon them. In this chapter, the various approaches to th identification and tracking of quality costs are discussed and its objec tives are:

- To understand the importance of quality costing
- To introduce the main approaches to quality cost measurement
- To gain an appreciation of the key issues relevant to the collection o quality costs
- To consider how best to present quality cost information.

What are quality costs?

Quality costs are those incurred by an organization in ensuring that th product or service it provides conforms to customer requirements Therefore quality costs are the sum of that spent in ensuring tha customer requirements are met plus that wasted through failing to achieve the desired level of quality. They can arise from any activity that impinges on the quality of the product or service, be it a cos associated with providing the end product/service or a support activ ity. Such costs can be influenced by activities internal to the organiza tion, such as:

- Market intelligence
- Design
- Production
- Delivery
- Maintenance.

or by external groups such as suppliers and customers.

The case for quality costing

Quality costing is one of several techniques available to assist companies with the attainment of quality. It is important primarily because the costs of quality are large. In manufacturing industry they have been estimated to be between 5% and 25% of sales turnover, whereas in service industries they are considered to be even higher, at 30-40% of operating costs. In addition, up to 95% of this cost may be expended on appraisal and failure. In some service organizations it has been estimated that up to 60% of employee time is spent checking, rectifying and apologizing for errors.

It is indefensible that such costs, especially those that are avoidable, should be unmeasured and hence uncontrolled in any organization. Indeed, studies in many industries have shown that with management commitment, and a focus on continuous improvement, they can be halved in a 3–5-year period.

Many quality gurus are in support of the case for quality costing. Feigenbaum[1] believes that quality is as fundamental a way of managing the business as marketing or production. He advocates the direct management of quality costs, measuring them, controlling them and using them for strategic planning and budgeting. Juran[2] speaks of the 'gold in the mine' as the total avoidable costs of quality. He argues that the costs resulting from defects are a gold mine waiting to be excavated. Juran discusses the many in-built inefficiencies such as excess materials, allowances, excess starts and deliberate overproduction that are built into normal everyday practice for contingency purposes. Apart from their costs, these inefficiencies distort the basic values against which important judgements are made and, ironically, the more the base values are used, the more firmly entrenched and accepted the built-in inefficiencies become. This leads to unnecessary and avoidable cost penalties which make goods and services more expensive and, in turn, uncompetitive.

Although there is a good case for quality costing, it will not solve quality problems on its own. It must be used as part of a wider total quality management approach, where problem areas are identified and prioritized using quality cost data and improvement activity is initiated.

There are a number of approaches to categorizing and reporting quality costs. In this chapter the two most common approaches, known as the prevention-appraisal-failure (PAF) model and the process cost model (PCM), will be discussed.

The prevention–appraisal–failure model

Dr Armand V. Feigenbaum is attributed with having developed th concept of quality cost measurement while working in the Genera Electric Company in the 1950s. He proposed a system of 'Qualit Costs Reporting' whereby all relevant costs were categorized int four major areas:

1 Prevention costs (also known as the cost of conformance)
2 Appraisal costs
3 Internal failure costs (also known as the cost of non-conformance
4 External failure costs.

Prevention costs

These are costs incurred in preventing quality problems arising. Th costs of any action taken to investigate, prevent or reduce non-confor mities or defects and the cost of planning, introduction and main tenance of the quality system would be included in this category. I is reasonable to expect that expenditure here would reduce all othe quality costs.

Appraisal costs

These costs are incurred in assessing the conformance of the product o service to requirements. This category would include such things a receipt testing of goods and all inspection and testing during produc tion. It is arguable that appraisal costs are capable of reduction whe there is an emphasis on quality improvement.

Internal failure costs

These are the costs arising within the organization of the failure t achieve the quality specified (before transfer of ownership to the cus tomer). This cost is incurred because something was not done 'righ first time' and in theory it would disappear if there were no interna defects such as scrap, no need for rectification or redesign and n delays to the production process due to non-conformance and non conformance-generated shortages. Internal failure costs are most read ily identified and examined for quality cost reduction.

External failure costs

These are the costs arising outside the organization of failure to achiev the quality specified (after transfer of ownership to the customer) Again these costs would disappear if there were no external defects warranty claims, replacement costs, etc. Care must be exercised in interpreting this category as only a partial story is told. Quantification may not reflect the loss of customer goodwill or future loss of sales, both of which are examples of external failure costs. This cost has traditionally been reduced by high levels of checking, which increases internal failure costs as more failures are detected in-house.

Prevention is the only justifiable element of the quality costs listed above (although appraisal may be justified in industries where safety is an important issue). Since the costs incurred by appraisal activities and failure are totally avoidable (in theory at least) they are jointly described as the cost of poor quality (COPQ). Typical COPQ figures of 15–40% of the cost of sales have been recorded in a number of companies. Thus these avoidable costs are usually greater than double their profit margins.

One may ask why action to reduce the cost of poor quality is not the key priority of every company. The answer is that many of the contributions to quality costs have never been identified. Examples include inadequate marketing, clerical errors, poor plant layout, long design cycle times, etc. In other cases many of the costs that have been identified are often considered inevitable. Examples include over-ordering of material to allow for scrap, redrawing, drawing office queries, excess inventory levels, penalties for late deliveries, etc. If such costs are to be reduced then they must be identified and reported to management. In other words, some method of classification and collection is required.

Elements within quality cost categories

Each category can be divided into distinct cost elements. The following list appears in BS 6143: Part 2: 1990:[3]

Prevention Costs

- Quality planning
- Design and development of quality measurement and test equipment
- Quality review and verification of design
- Calibration and maintenance of quality measurement and test equipment
- Calibration and maintenance of production equipment used to measure quality
- Supplier assurance
- Quality training
- Quality auditing
- Acquisition, analysis and reporting of quality data
- Quality-improvement programmes.

Appraisal costs

- Pre-production verification
- Receiving inspection
- Laboratory acceptance testing
- Inspection and testing
- Inspection and test equipment
- Materials consumed during inspection and testing
- Analysis and reporting of test and inspection results
- Field performance testing
- Approvals and endorsements
- Stock evaluation
- Record storage.

Internal failure costs

- Scrap
- Replacement, rework and repair
- Troubleshooting or defect/failure analysis
- Re-inspection and retesting
- Fault of sub-contractor
- Modification permits and concessions
- Downgrading
- Downtime.

External failure costs

- Complaints
- Warranty claims
- Products rejected and returned
- Concessions
- Loss of sales
- Recall costs
- Product liability.

Exercise 6.1

A manufacturer has collected the following quality cost data. Assign the figures to the four cost categories and present your findings as a percentage of the total.

	£
Scrap	75 000
Rework and repair	80 000
Warranty costs	21 000
Downgrading	31 000
Final inspection	65 000
Returned material rectification	55 000
In-process inspection	33 000
Goods inwards inspection	27 000
Retesting	28 000
Customer complaints handling	19 000
Quality planning	19 000
Product liability claims	72 000
Vendor quality assurance	22 000
Quality system audits	13 000
Calibration and maintenance of test equipment	7 000
Quality improvement programme	46 000
Quality training	8 000

Answer to Exercise 6.1

	£
Prevention costs	
Quality planning	19 000
Vendor quality assurance	22 000
Quality system audits	13 000
Calibration and maintenance of test equipment	7 000
Quality improvement programme	46 000
Quality training	8 000
	115 000
Total costs =	621 000

Prevention costs = 18.5% of total quality costs

Appraisal costs	
Final inspection	65 000
In-process inspection	33 000
Goods inwards inspection	27 000
	125 000

Appraisal costs = 20.1% of total quality costs

Internal failure costs	
Scrap	75 000
Rework and repair	80 000
Downgrading	31 000
Retesting	28 000
	214 000

Internal failure costs = 34.5% of total quality costs

External failure costs	£
Warranty costs	21 000
Returned material rectification	55 000
Customer complaints handling	19 000
Product liability claims	72 000
	167 000

External failure costs = 26.9% of total quality costs

The process cost model

A new approach to quality costing, complementary to the PAF mode is called the 'process cost model'. This approach, which is suitable fo both manufacturing and service industries, is based upon the concep that each activity in an organization forms part of a process and tha the cost model should reflect the total cost of each process.

A process can be considered as any activity that converts inputs int more valuable outputs, utilizing resources and being subject to parti cular controls (see Figure 6.1). Process inputs come from suppliers an consist of data and/or materials that are transformed by the process t create outputs. Process outputs are delivered to customers and ma include that which conforms or does not conform to requirements, an may also consist of process information and waste. Controls consist o inputs that define, regulate and influence the process (e.g. policies strategies, procedures, methods, standards and legislation). Resource are contributing factors which are not transformed to become the out put. These may include people, equipment, material, accommodatio and internal/external environmental requirements.

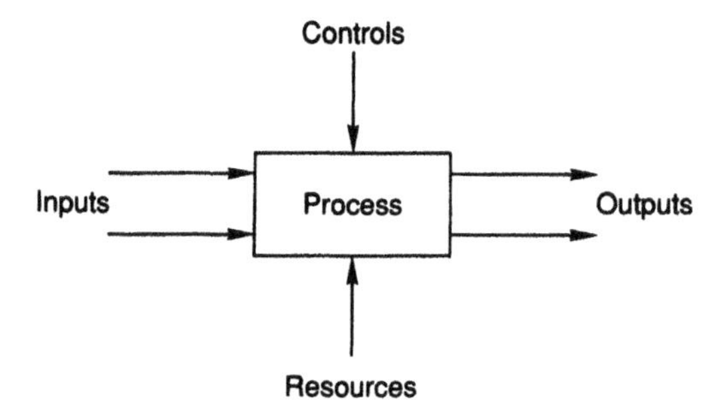

Figure 6.1 *The basic process model*

The PCM can be applied to any process or sub-process that support the critical processes associated with the delivery of a product o service to the customer. It is based on the assumption that the only partitioning which is valid when considering the process cost is tha between the costs of conformance (COC) to requirements and the costs of non-conformance (CONC). Care should be taken not to confuse these with similar terms used within the context of the PAF model. The COC consists of the intrinsic cost of providing products or services to declared standards by a given, specified process, in a fully effective manner. It is the minimum cost for the process as specified. The CONC is the cost of inefficiency within the specified process such as wasted time, materials and resources, associated with the process. It consists of unplanned and non-essential process costs arising from unsatisfactory inputs, errors made and rejected outputs. There are therefore two ways of reducing such costs:

- Change the process to increase added value, reduce costs and prevent failure
- Establish the cause of failure to conform and take preventative action.

Constructing the PCM

To use the process cost model it is necessary to introduce a system to analyse the process and separate the costs due to conformance from those due to non-conformance. This information can be used to identify aspects of the process that require improvement while helping in the allocation of key resources and major priorities. Guidance on use can be found in BS6143: Part 1: 1992.[4]

A PCM can be constructed for any process or sub-process, or to monitor the overall cost of a department. The preparation of the model consists of seven stages and begins with the construction of a block diagram to identify all the elements of the process:

1. Name the process (e.g. provision of hotel restaurant facilities) and assign a process owner (see Figure 6.2(a))
2. Define process boundaries
3. Identify outputs and customers (see Figure 6.2(b)) and identify inputs and suppliers (see Figure 6.2(c))
4. Identify controls, resources and source (see Figure 6.2(d))
5. Flowchart the process to identify key activities
6. Identify the cost elements associated with each activity. For each cost element, list it as a COC and/or CONC and identify the source of the data. The latter may be an actual cost as used by the finance department or a synthetic cost derived from available data on a clearly established basis. Table 6.1 shows an example of a typical cost model
7. Produce a process cost report assigning the costs against the agreed cost categories such as people, equipment, materials and the environment. The means of calculation for each element of cost may also be specified.

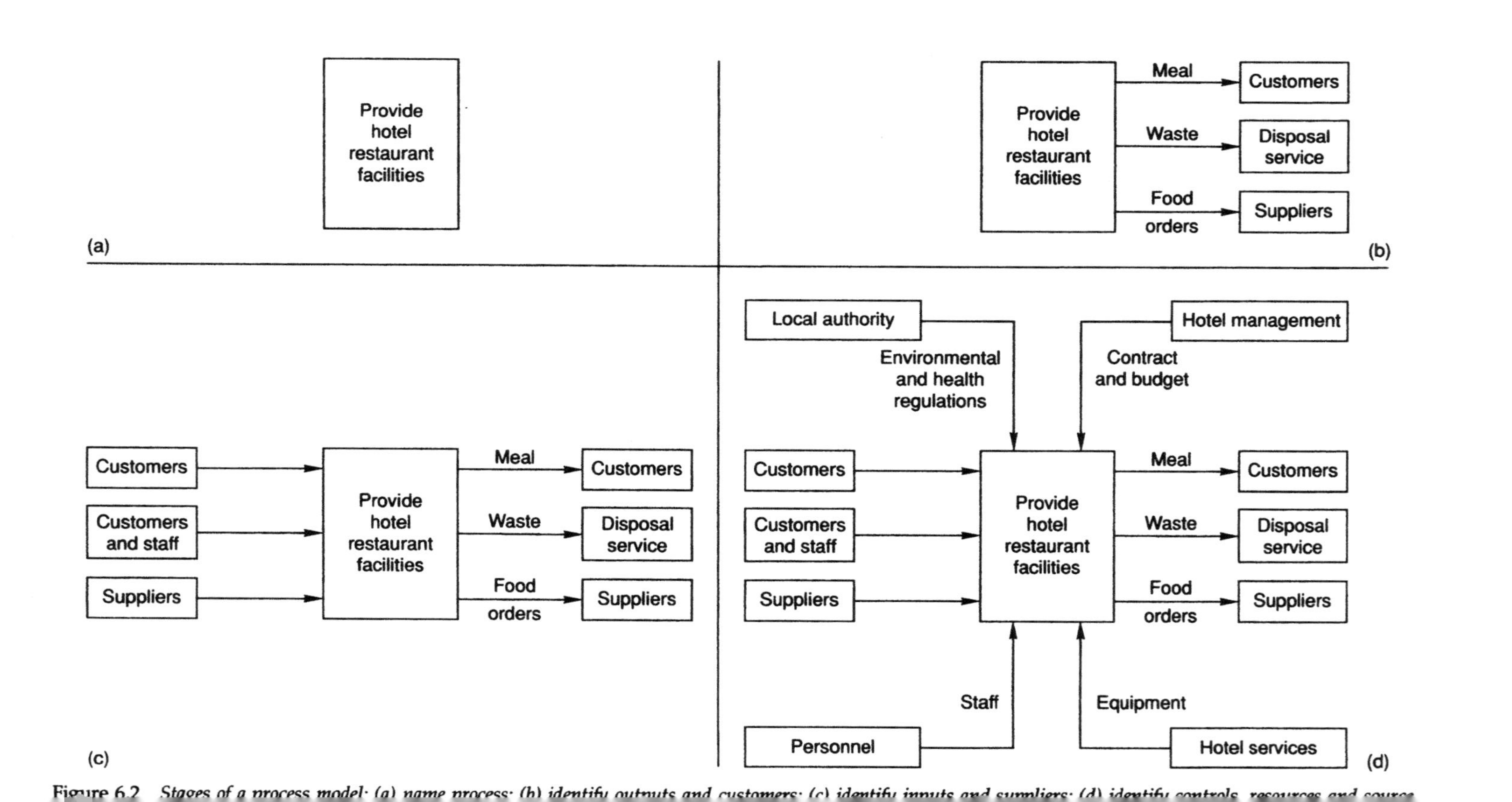

Figure 6.2 *Stages of a process model: (a) name process; (b) identify outputs and customers; (c) identify inputs and suppliers; (d) identify controls, resources and source*

Table 6.1 Identification of costs for key activities

Key activity	*Process costs relating to provision of hotel restaurant facilities* *Cost of conformance*	 *Cost of non-conformance*
Plan menu and order food	Cost of labour to produce planned menu and to order correctly	Cost of waste food due to overplanning Cost of offering a more expensive alternative due to underplanning
Prepare and present food	Material, labour and energy costs to produce planned food	Cost of waste food and labour etc. due to incorrect preparation
Serve food	Cost of serving food to customers within standard times	Cost of waste due to delays or incorrect servings
Handle complaints		Cost of investigations and rectifications
Clean restaurant	Cost of material and labour to clean to specified standards	Recleaning costs Cost of prohibition/ improvement notices
Train staff	Cost of planned training	Cost of cancellations Cost of ineffective training

The cost model should be constructed and monitored by the person who has full responsibility for, and authority over, the process (the process owner), perhaps in conjunction with a quality improvement team. The information in the report can be used to define priorities for improvement and to monitor the resultant cost changes. The effectiveness of the model should be also be continually reviewed to ensure its effectiveness.

The need for a process should always be reviewed before constructing a cost model. For example, some processes may only exist because of non-conformance elsewhere (e.g. a process for correcting errors) and could be removed if that non-conformance is removed. Once such decisions have been made, the balance of COC and CONC can be used to determine the need for process redesign or improvement.

Exercise 6.2

The human resource department of a company has analysed its key activities and identified the planned and unplanned costs. Prepare a process cost report for the organization and calculate the percentage of CONC to COC.

People satisfaction surveys
Total cost= 800 hours × 15/hour = 12 000

Staff turnover
Total cost= 420 hours 15/hour + 22 000 paid in lieu =28 300

Industrial action
Total cost = 370 hours × 15/hour = 5550

Recruitment costs
Cost of satisfying requirements = 75 hours × 15/hour = 1125
Cost of inadequate recruitment = 12 500 extra training

Training costs
Cost of training = 11 5000
Cost of non-attendance = 5500
Floor space and services total cost = 80 000
Training materials invoiced cost = 14 000
Hire of specialist equipment invoiced cost = 2500

Answer to Exercise 6.2

Process COC	Process CONC	Cost (£)
People		
Satisfaction surveys		12 000
	Staff turnover	28 300
	Industrial action	5 550
Recruitment costs		1 125
	Inadequate recruitment	12 500
Cost of training		115 000
	Non-attendance	5 500
Environment		
Floor space, etc.		80 000
Materials		
Training materials		14 000
Equipment		
Hire charges		2 500

Total costs = 276 475
COC = 224 625
CONC = 51 850 (23% of COC)

Advantages of the PCM

Process cost modelling has the following advantages:

- It aids understanding of the state of a process and how it is functioning
- It supports the requirement for effective process management and control
- It generates commitment for investment in process improvement

- It promotes the involvement of everyone in the organization in reducing costs and continuous improvement.

The PAF versus PCM approach

One of the main difficulties of the PAF approach relates to the categorization of costs as prevention, appraisal and failure costs. Certain activities, such as design reviews, which may be considered a prevention cost, can also be classified as appraisal costs. This can cause confusion and arguments which detract from the main message that a cost exists that should ultimately be reduced.

Despite such problems, there may be occasions when it is desirable to link the traditional PAF approach with the PCM methodology. Those organizations that have successfully implemented, and gained acceptance of, the PAF approach may not wish to adopt an entirely new approach. In such circumstances, the COC may initially be considered as that pertaining to the PAF approach, where prevention and appraisal comprise the COC and failure costs comprise the CONC. The COC indicates the cost of conforming to requirements and not necessarily an efficient or necessary process. As a consequence, such costs offer an opportunity for reduction and should be considered as part of continuous improvement activities.

Collection of quality costs

If data is to be collected the cost categories must be identified correctly in the environment under investigation. Guidelines to the collection and analysis of quality costs are given in BS 6143. Unfortunately, due to the differences in accounting procedures within organizations, no set rules are available for establishing and controlling quality costs. So although such publications give guidance on the general principles, the procedures and make-up will vary and organizations should develop a system that meets their own needs.

An organization must decide for itself the most appropriate method for collecting and categorizing quality costs. In developing an approach, a thorough understanding of the organization is required, so ensuring that the system is applicable across the entire operation and that the information is useful. The influence of existing approaches to the classification and allocation of costs must also be considered. For example, the manner by which overheads will be apportioned must be determined. In some cases quality-related costs would not normally attract overheads, whereas in others they would be treated as direct costs and would attract a proportion of overheads. Failure to clarify this and similar issues could lead to a distortion of the facts and a lack of acceptance of the information itself.

Once the relevant categories have been identified the availabl sources of information within the company can be examined. Dat may be obtained from many sources such as:

- Payroll analysis
- Expense reports
- Field repair and replacement reports
- Inspection records
- Travel expense claims
- Time sheets.

However, not all the information required will be quoted in a suitabl format and so before any progress can be made it is important tha clear objectives are defined for the use to which the data will be pu This will influence the method of collection as well as the data to b collected, since if estimates are sufficient, existing company system may provide adequate information. If, on the other hand, accurat data are needed then special reporting systems may be required.

Most quality costs are not reflected in the charts of accounts o existing cost accounting systems. Hence the accountant will be unabl to provide all the figures required with respect to the various cos categories. Useful information on scrap and loss is normally availabl and its value can be evaluated quite easily. The figures for those qualit cost categories which coincide with accounts that are elements o departmental budgets (e.g. time spent on test activities) are also avail able. However, items such as rework and repair are more difficult t evaluate as the costs associated with these can arise in a number o different ways.

As previously stated, the approach used to supply missing figure depends on what is the purpose of quantifying quality costs in the firs place. If the main purpose is to identify opportunities for reducing th costs, the missing figures can be supplied by searching existin accounting figures, analysing the basic data input to the accountin system and making estimates where no effective recording system i available. This method is attractive to the busy manager as it require no special systems. However, the basis upon which estimates are mad must be clearly understood and agreed by everyone to ensure consis tency and acceptance.

A view held by some practitioners is that accuracy is an importan criterion as it has direct influence on the amount of work required an the credibility of the outcome. They suggest the costs should be accu rate enough to be credible even to those whose efficiency or perfor mance is called into question by the resulting report. Where this i important, the help of the accounts department should be sought i compiling the report. Indeed, if the main purpose of collecting an reporting costs is to set targets and monitor results then the approac used may require the allocation of account codes within cost centres fo each element or quality cost category. These would enable everyone involved to log their time etc. and would enable costs to be traced to individual areas, including suppliers and sub-contractors.

Such systems can be used to highlight everyone's contribution to profitability. However, the actual practicalities of setting up and operating these systems can prove quite difficult if, for example, the cost data cross departmental lines. Some accounting systems are not designed such that cost collection exercises crossing conventional boundaries can easily be accomplished. Departments and different locations may even have to develop their own detailed systems.

Reporting quality costs

Effective quality cost control depends on good cost reporting. Any report must be appropriate to the needs of the user (e.g. provide information which will identify problem areas, help set targets and monitor results). Detail should be shown where necessary, trends plotted, comparative ratios calculated and important points highlighted.

Good standards of reporting are essential if reports are to make an impact and provoke action. Management require data that facilitates decision making and so it is important that costs are not obscured by unnecessary information. This makes the picture less clear and provides an excuse to defer a decision.

To have sufficient impact, the quality cost report should be presented in a format similar to existing management accounts, and should be supported by financial ratios and trend analysis. Where practicable, the same frequency and periods as for other accounts reports should be used to allow realistic comparisons to be made. Typical examples of reports based on the PAF and PCM approaches are shown in Tables 6.2 and 6.3, respectively.

There are a number of criteria which a good system should meet. It should:

- Identify the areas of expense being reported
- Show actual expenditure compared with that planned
- Facilitate the comparison of benefits with the price being paid for them
- Attempt to measure every category of cost to be certain that an adequate balance has been achieved and that important areas are not omitted
- Attempt to represent the current situation and not a previous one, so speed is essential
- Highlight important trends so that problems can be identified and improvements monitored
- Segregate costs in detailed areas wherever possible

Table 6.2 PAF cost report

Current period				*Year to date*		
Act	*Bud*	*Var*		*Act*	*Bud*	*Var*
			Prevention			
			Quality planning			
			Quality improvement			
			Quality auditing			
			Calibration			
			Others:			
			Appraisal			
			Production testing			
			Receipt inspection			
			Stage inspection			
			Final inspection			
			Others:			
			Internal failure			
			Nonconformance			
			Cost of excess inventory			
			Waste heat and light			
			Downtime			
			Others:			
			External failure			
			Transport premium			
			Loss due to late delivery			
			Warranty claims			
			Complaint handling			
			Others:			

- Create a form of early-warning system, i.e. be able to identify wher rapid corrective actions can be taken to prevent or limit avoidabl expenditure
- Establish cost-improvement objectives to encourage management t challenge existing costs
- Relate avoidable costs to profit (or budget) since a reduction in tota costs can make a substantial impact on profitability.

The presentation of quality costs in their respective categorie together with the total quality cost reveals their relative magnitude However, total quality costs on their own do not give the full pictur of what is happening to quality in relation to other business activity. I is useful to compare them, or a cost category, with other activities in th business. There are a variety of ways of doing this, depending o whether the basic purpose is to embark on an improvement pro gramme or to compare and monitor costs. The most popular method are ranking, ratios and trend graphs.

The simplest analysis involves recording the data in descendin order of magnitude against each cost element. This then gives som indication as to which elements may be examined first to reduce qual ity costs effectively.

Table 6.3 PCM cost report

Process conformance	Cost			*Process nonconformance*	Cost		
	Act	*Syn*	*$*		*Act*	*Syn*	*$*
People							
Hours taken to produce menu	*						
				Cost of alternatives	*		
Hours taken to prepare food	*						
				Hours spent on wasted food	*		
Hours serving food	*						
				Hours investigating complaints		*	
Training	*						
				Cancellations		*	
				Cost or retraining	*		
Equipment							
Cost of capital equipment	*						
Environment							
Cost of heat and light and floor space	*						
Hours spent cleaning	*						
				Hours spent recleaning	*		
				Cost of rectifying env & health problems		*	
Materials and methods							
Purchased food	*						
				Waste food	*		
Cleaning materials	*						

A popular ratio compares total quality cost with the value of sales, or operating budget, as it makes the job of grasping their significance much easier. When using sales it is important to recognize that sales include a profit element which can vary from period to period and so care must be taken when interpreting results. Quality costs can also be compared with profit or margins, which often encourages management to take notice of them. Another popular ratio compares total quality costs to value added.

Any ratio may be taken to establish the percentage contribution made by an item to achieving a particular condition. Quality cost interrelationships, using the PAF categories, can be examined and this often shows that even though appraisal costs are budgeted fo failure costs are much higher (e.g. 30% versus 65%). It will often b found that prevention costs are usually very low in proportion to th total (e.g. 5%), and such information normally encourages managemen to look at the possibility of increasing prevention activities. Other ratic often track the percentage contribution of a cost element to a cos category (for example, the percentage of failure costs due to scra and rework).

The particular ratio used will depend upon the type of business an the aspect of financial performance of interest. For example, labou intensive services would be interested in improving personnel utiliza tion and so would use a labour-based ratio such as internal failure cos versus direct labour costs. However, a manufacturing organizatio investing in automation would not use this ratio since employee reduc tions would change the ratio regardless of any reduction in failu costs.

All those who are in a position to influence costs in future period should receive a copy of the report. Some companies believe that suc reports are confidential and should only be circulated to senior mar agement. Others believe that all those who either contribute to th report or who are responsible for the costs should get a copy. Th latter view emphasizes that visibility is important if quality is to b improved and costs reduced.

Exercise 6.3

Design a quality cost reporting format suitable for use in your ow organization. Discuss the information collection and analysis require ments with your management accountant and prepare an initial qualit cost report.

Uses of quality costs

The potential uses of the information contained in a quality cost repor are numerous but may be grouped into three broad categories. Firs quality costs may be used to promote quality as a business paramete Second, they give rise to performance measures. Third, they provid the means for planning and controlling future quality costs.

Quality as a business parameter

The first use – promoting quality as a business parameter – is usuall expressed as gaining the attention of higher management by using th language of finance. This allows quality to be treated as a busines parameter like any other activity. Costs can also be used to illustrat that it is not only the quality department who are involved in quality, since everyone's work can impinge on the quality of the product.

It may be argued that quality problems are already known in the organization and so it is pointless spending time and money converting them into costs. This may be true in the case of certain elements such as scrap and rework, but most managers will have no idea of the size of these costs and will be alarmed when they are collected. Many executives have now realized the need for continual improvements in quality, and this has prompted quality cost investigations which can reveal the magnitude of the problems.

Performance measures

The second use – giving rise to performance measures – is usually expressed as absolute or relative costs, i.e. indexes or ratios to other business costs. These cost measures are used for three main purposes:

1 Comparison with other parts of the business or with other businesses
2 Decision making
3 Motivation

Comparisons between sets of quality cost data enables current performance to be measured against previous costs. It is important that both sets of data should be arrived at on the same basis, and for this reason most companies compare only internal data and do not use the data for comparisons between companies. Apart from using quality costs for comparative purposes, they can also be used to learn more about the economics of investment in prevention activities.

The use of quality costs for decision making refers to its use in helping managers to identify major opportunities for cost reduction. This means that realistic estimates must be available and costs must be capable of relatively quick collection (i.e. in weeks or months rather than years). The costs collected can then be used to assist in identifying projects for improvement. Techniques such as Pareto Analysis are often used to determine which projects to tackle first.

Costs focus attention on the chronic problems for which compensations have been built into the system. The argument is that such problems are readily picked up by other means but that the results of such problems are built into the base values against which judgements are made. A typical example is the allowance for scrap built into standard costs.

The use of quality costs for motivational purposes include displays to shopfloor workers of scrap costs arising within their centre and to middle managers for their departments' contribution to total quality costs. Once it has been recognized that costs can be reduced, new targets can be set.

Planning and controlling costs

The use of quality costs as a means of planning and controlling cost include establishing quality cost budgets for the purpose of cost con trol. Quality costs can also be used to follow the progress of cost reduction programmes. Interest will be shown in a programme tha starts to show encouraging results. It is therefore necessary for th cost figures to move from estimates to historical returns. If the esti mates are realistic this shift from estimates should not show a sizeabl difference in the figures. However, as the company becomes mor aware of the requirements for good quality and the scope of qualit costs, the reported values will change.

Quality cost reports can also help hold the gains. Any improvemen in quality level and reduction in quality costs must be maintained to b of long-term benefit. New standards can be set at the improved level and it will then be necessary that the gains are being held or furthe improved upon.

Quality cost reporting also improves communications with uppe management. Because communication is via a common language, dif ficulties in relating quality to other company activities can be overcom and quality activities can be better understood. Using financial termi nology also keeps quality matters continuously in the minds of senio management, just as sales and manufacturing accounts are usuall recognized and receive attention.

Quality cost reports are extremely useful. However, reporting an analysis will not improve quality on its own. The report only indicate those areas where effort must be placed if problems are to be resolved Other methods must then be used to investigate the causes and solv the problems. This is why the collection of quality cost data is often tool in a total quality programme. Indeed, taken overall, quality improvement projects appear to offer the quickest route to usefu exploitation of cost data. The most valuable use is to help companie to decide how, when and where to invest in prevention activities.

Conclusions

Quality costs are a useful management tool and can be used for bot planning and control purposes. However, there are many complexitie and difficulties to be overcome if the financial aspects of quality are t be raised from the level of *ad-hoc* cost reports and occasional cost reduction exercises to the level enjoyed by other major business para meters.

Deciding which activities should be included under the quality cos umbrella is by no means straightforward and there are many grey areas. Some quality practitioners tend to include costs that are difficul to justify as quality-related (e.g. the preparation of engineering and administrative systems and procedures, document and drawing controls, etc.). Whether such costs should be included is open to debate, and in many companies such costs are excluded to avoid arguments over cost elements.

Quality costing does have limitations in its use. It will not in itself, energize managers into ways to reduce the costs. An organized approach, whereby promising improvement projects are identified, the causes diagnosed and remedies put into place, is required and this is why quality costing should be used as part of a continuous improvement programme. Otherwise quality costs will just be used for control, and as long as the costs are no worse than usual, no alarm signals to initiate action will be given and the costs will remain at the same level.

Relatively few companies have a full quality cost reporting system. If there is a desire to raise the profile of quality in the business then linking it with finance is likely to excite most people. This will ensure that the subject is taken seriously and regularly discussed. The task is not easy, and there may be opposition and obscuration of the data but it is worth while. Quality cost reporting is a vehicle on which a greater quality awareness can be introduced. This is vital if an organization is to remain profitable in an increasingly competitive market.

Summary points

- Quality costs relate to expenditure on ensuring conformance to customer requirements.
- Quality costs are large, having been estimated to be between 5% and 25% of sales turnover in the manufacturing sector and 30-40% of operating costs in the service sector.
- The PAF model categorizes quality costs with four major areas.
- Major reductions in quality costs will only be achieved through management commitment and an investment in quality improvement.

References

1 Feigenbaum, A V. (1983), *Total Quality Control*, New York: McGraw-Hill.
2 Juran, J. M. (1951), *Quality Control Handbook*, New York: McGraw-Hill.
3 BS 6143: Part 2: 1990: Guide to the economics of quality: Part 2. Prevention, appraisal and failure model, British Standards Institution, London.
4 BS 6143: Part 1: 1992: Guide to the economics of quality: Part 1. Process cost model, British Standards Institution, London.

7 The organization of quality

Introduction

Total quality management is an approach to management whic focuses on giving value to customers by building excellence int every aspect of the organization. This can be achieved by creating a environment which allows and encourages all employees to contribut to the organization and by developing the skills that enable them t study scientifically and constantly improve every process by whic work is accomplished.

In all organizations there are processes by which tasks are carrie out. There are processes of production, sales and distribution. There ar processes which combine market information with new technologies which in turn generate ideas for new products and services. Othe processes create and test these new products and services and translate them into production.

In studying these processes the continual aim is to execute then better and to provide customers with products and services of ever increasing value at lowering cost. Certain guidelines in respect o achieving this objective have become apparent:

1. *The quality organization must listen to customers and help to identify an articulate their needs.* This involves knowing in detail the wor your customer is involved with, how a customer uses your products, what problems a customer has. It is also important to know what your non-customer base is, as people who are not yet customers can provide important feedback on your product or service.
2. *Products and services which satisfy the customer result from well-planne and executed systems and processes.*
3. *The organization's vision, values, systems and processes must be consistent with each other.* As an organization has thousands of processe the result of non-uniformity is waste and frustration. Engineers, fo example, may design a product that production do not have th capability to make. Purchasing many buy materials that productio cannot use. The work of one aspect of a process must be complementary to the next step.
4. *Everyone in the organization must work together.* People normally form bonds in relation to profession, function or rank. To enable all systems to work in a coordinated manner a spirit of teamwork should exist that is more common than these traditional bonds.

5 *Teamwork in a quality organization must be based on commitment to the customer and to constant improvement.*
6 *In a quality organization, everyone must know their job.* This involves:
- Employees understanding their personal role within the context of the organizational whole (i.e. their standing in the sequence of activities; the relationship between their activity and the final product or service to the end user or consumer)
- Employees knowing their internal customers
- Employees possessing the skills necessary to perform tasks related to their activities.

7 *A quality organization uses data and a scientific approach to plan work, solve problems, make decisions and pursue improvements.* The most effective methods of accomplishing work are determined by the gathering and analysis of data relevant to that work.

The objectives of this chapter are to enable the manager to:

- Determine organization structure to support the improvement initiative
- Provide advice and encouragement to individuals in defining their roles and responsibilities
- Contribute to developing ownership and commitment to quality within individuals by ensuring that their development needs are accurately assessed against performance standards relevant to their roles and responsibilities
- Determine the specifications to secure quality people
- Assess and select candidates against team and organization requirements
- Allocate work and evaluate individuals against objectives
- Provide feedback to individuals on their performance
- Develop awareness of reward and recognition schemes.

The changing concept of organizations

Total quality management is a process of change. It is a managerial process and requires a clear statement of purpose, a planned strategy of implementation and a means of evaluation, feedback and follow-up. However, for many organizations the development of a culture which nourishes the improvement efforts of every group and individual represents a marked change in philosophy. Corporate culture has dominated much management thinking in recent years, with many writers examining a cultural perspective on management and organizational issues.

In general, definitions of culture tend to deal primarily with the way we are or the way we think. Marguiles and Raia[1] define culture thus: 'The commonly shared beliefs, values and characteristic patterns of behaviour that exist within an organization'.

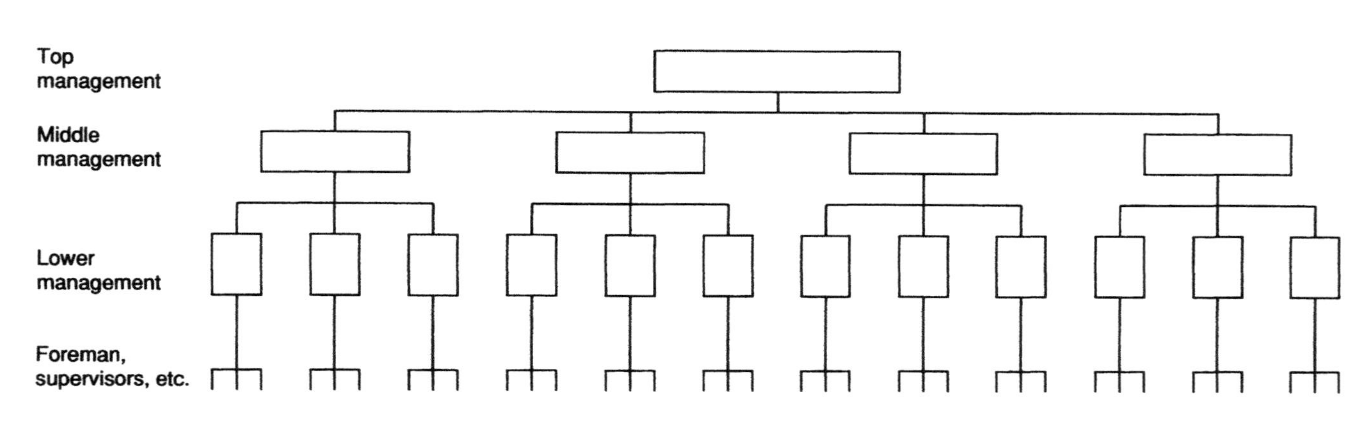

Figure 7.1 *Hierarchical structure*

Bower[2] has given the most commonly accepted definition: 'Culture is the way we do things around here.' The way most managers 'do things around here' is encapsulated in the traditional hierarchical structure of business enterprise: the chain of command or management by control (see Figure 7.1). This classic organization chart was developed in the 1840s. Devised when businesses needed, for the first time, to manage mass production and wider distribution, it depicts the separation and decentralization of functions. It also describes the downward path of a system of control and accountability. Each employee is governed and evaluated by a set of numerical objectives, performance standards or work quotas. Management by control is simple and logical – the sum of the accomplished objectives at one level will fulfil the objectives of the person immediately above.

However, as management by control is a system of controls, the rewarded accomplishments are, by necessity, measurable and short term. Priority is given to the immediate and measurable and not on the intangible long term. This short-term focus can also lead to an inward focus. If meeting the short-term goal is the objective then individual success and the success of the system of control is determined by how well *imposed* goals are achieved.

What, for example, if measurable goals are impractical? Employees may 'play the game' of fabricating performance, managers of different departments may adopt an over-defensive attitude. Fear can easily become the prime motivator. As already indicated, the purpose of a quality organization is not accountability and control in these terms. Such a rigid approach does not:

- Depict the interdependence of various functional areas
- Describe the organization as a flow of processes
- Emphasize group accountability rather than individual
- Refer to customers.

The organization as a system

If a quality organization is one which meets or exceeds customer expectations then each internal process must work well and interact with those processes which precede and follow it. The organization as a system of interrelated activities was developed by Deming[3] (see Figure 7.2). The focus here is on the processes by which work gets carried out and not on a hierarchy of individual accountability. It emphasizes:

- The interdependency of organizational processes
- The importance of the customer or consumer
- The impact of customer feedback (consumer research)
- Continuous improvement based on customer feedback

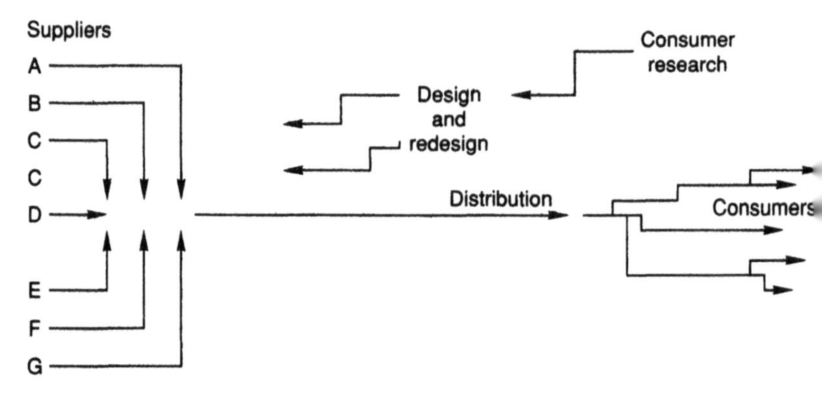

Figure 7.2 *The organization as a system*

- The importance of suppliers
- The network of internal supplier/customer relationships.

Managing change

Total quality is a process of change and no single change is mor important than that of employees' attitudes. A shift in managemen style is necessary to allow 'change' to happen. If we consider th organization as a social system, a loose network of small groups o people, then it is inevitable that certain groupings form, certain peopl emerge as leaders of groups and these groups then develop their ow 'rules'. These rules can determine the pace of work, the nature o relationships with other groups and the degree of acceptance or resis tance to change.

People do not necessarily resist change, but they may resist actuall being changed. In planning or leading a change initiative the logica and rational approach can have limited effect:

- Top-down orders can be ignored or interpreted in many differen ways
- Coercive methods of change reinforce the chain of comman approach.

People need to feel included in the decision to change and be aware o the rationale for change. Individual reservations, fears and concern need to be accounted for. The process of change, as Kubler-Ross[4] ha identified, involves seeing transition in perspective:

Denial	This crisis shall pass away
Anger	Why should I change!
Bargaining	Can we work out a compromise?
Fear	I don't know if I can cope
Resignation	OK, let's carry on.

The 'undecided' typically represents the majority response to change. It is unrealistic to attempt to convert all employees to the need for a proposed change. However, if a sufficient number of influential people within this 'critical mass' can be 'recruited' then the momentum for change gathers both credibility and interest. Influential people could be division managers or a number of committed people in any particular area. Without the support of management a proposed change has little chance of success.

Dealing with resistance to change

Change is often resisted due to:

- Fear of economic loss (more work for same pay, loss of overtime)
- Change in perceived security (possible layoffs, difficulty in learning new skills or routines)
- Conditions of work (change in hours, procedures)
- Job satisfaction (less challenge, closer supervision, reduction in authority)
- Annoyance with way change was handled (misunderstanding of reasons for change, change made too quickly)
- Cultural beliefs (change not consistent with tradition, mistrust of management).

Reducing resistance

The most popular approaches to deal with resistance to change include education and communication, participation and involvement, facilitation and support.

Education and communication

This method deals with the lack of information or accurate information about process changes. Educating employees and helping them to understand the logic and need for change is recommended.

Participation and involvement

The objective here is to allow the initiators of change the opportunity to increase information levels so as to create more effective change while encouraging those who might resist to become involved.

Facilitation and support

Being supportive in a time of adjustment can include providing emotional support, listening, supplying training for acquiring new skills.

Management of culture change

If culture is interpreted as 'the way things are done around here', then
it is the result of the common, on-the-job experience of the majority of
employees. To determine exactly how 'things are done around here'
consideration should be given to determining:

- What are the unwritten rules of the organization?
- What are the organization's taboos, clubs, cliques, myths?
- How does working 'here' differ from working in another organiza-
tion?

Managers at all levels, therefore, should ask:

- How does the individual employee feel about working here?
- How do employees feel about the company?
- How do employees feel about their respective work groups?
- How do employees feel about their individual jobs?

Employees who feel relatively good about their jobs, colleagues and the
organization itself are more likely to have a positive impact on quality
improvement activity (see Figure 7.3).

In order to come to terms with and understand the 'culture' of an
organization managers must begin to learn from employees. Appraisal
interviews, informal meetings, team product meetings, questionnaires
attitude surveys – all these are mechanisms capable of generating
information which help to identify:

- What quality problems are being experienced?
- What prevents pride of work?
- What prevents teamwork?
- What would help employees to feel more a part of the company?

If an individual, or group of individuals, is involved in a creative or
innovation activity with regard to implementing change they should
not be operating in isolation. A support network should be established
whereby one project team, for example, is facilitated by a team of
managers who can give support and advice. This process of reinforcement is extended by having more than one project team operating at
any one time. Different teams can benefit from each other – training
and technical assistance can be shared.

The establishment of such a well-coordinated network of activity can
help to facilitate intended change by developing a sense of common
purpose.

- Am I capable?
- Am I proud of the work I do?
- Is the work worth while?
- How do I feel about my work group?
- Is there a sense of teamwork?
- Is there trust and loyalty in the group?
- Is there sufficient collaboration?

- Am I made aware of organization development which directly affects me?
- Am I part of a team?

Figure 7.3

Managing the quality transformation

Senior management leadership

In the process of transformation to a quality organization senior managers must become leaders and teachers of quality. As a group they form the steering committee of the change process. Given that they develop strategies and plans, select targets and priorities, a more participative approach should be adopted in that they learn to see themselves as suppliers to a variety of internal customers whose expectations managers will learn to identify, meet and exceed. The organization's mission, values and strategic direction and the ways in which it achieves them should reflect the fact that:

- Feedback from customer and suppliers
- Feedback from employees
- Data on performance of competitors

have been considered. The process of formulating strategy includes the identification of specific inputs required and the areas to be addressed (for example, current market trends, changes in customer requirements). The development of a medium- to long-term strategic plan enables annual operating plans to be developed which address:

- Specific priorities
- Action plans
- Capability
- Training capacity
- Resource limitations
- Market trends

Mechanisms need to be developed which will help to identify clearly the needs and expectations of existing/potential customers. The use of surveys, questionnaires and personal visits will ensure that relevant information is gathered. Consideration should be given to involving customers as regards new product development and feedback on possible strategic development.

It is also beneficial to receive feedback from employees in terms of how strategies which have been developed impact upon them. Team meetings and presentations are potential methods of receiving feedback and can be used as mechanisms for developing commitment of employees to the implementation and realization of strategies. It also provides an opportunity for employees to be made aware of how the strategy impinges upon their own activities and could be used to set objectives for individual employees. In establishing a specific strategy with associated planning senior management should consider the following:

1 Where will we start? In which part of the company will we establish our starting point? Where will we target our first efforts? What strategic issues should we consider in the selection of our first efforts? Should we, for example, select areas where there is high visibility? Greater possibilities for major gains? Receptive key personnel? Critical need? Greater chances of success? Congruence with other corporate plans? Natural leads-in to future expansion of the quality implementation effort?
2 Who will be the person coordinating the implementation in the targeted area? How will he or she be prepared for that responsibility? What kind of ongoing development will he or she receive?
3 What kind of preparation will we provide to managers, supervisors, key staff people and union representatives who work within the targeted areas?
4 What specific activities will we, the top managers, undertake in the targeted area? How will we deploy ourselves? How will we prepare ourselves for this involvement?
5 How will we involve the appropriate middle managers and supervisors? How can we help them to understand, support and lead this effort in their respective areas?
6 Who will provide technical assistance in each targeted area? How will they be trained? How much of their time will be made available to improvement efforts?
7 What will be monitored in the initial stages to enable us to improve on performance?
8 In which parts of the organization will we extend our efforts after the initial stage?

Case study

The objective of this case study is to illustrate the typical behaviours senior management can demonstrate in driving the organization towards total quality. A total quality initiative demands that senior management set very clear direction for quality practices throughout the organization.

The case study is based on Valpar Industrial Ltd, a medium-sized manufacturing company which makes 'python' tubing, the purpose of which is to ensure that cold beer and soft drinks maintain their required temperature when being pumped from storage barrels to dispenser taps on bar and restaurant counters.

Visible involvement in leading quality management

- Communication through joint meetings
- Management involvement in quality action teams and project management teams. Fixed schedule for team meetings
- Management involved in the provision of training
- Open-door policy
- Clear quality policy communicated to all employees.

Management development of consistent total quality culture

- Total quality structure in place, with clear roles and responsibilities for individuals/teams involved
- Quality always on agenda of monthly sales and production meetings
- Job descriptions include quality obligations
- Monitoring of levels of awareness of employees
- Company goals and objectives are clear and communicated to all employees
- Annual appraisal considers commitment/achievement relevant to total quality initiative.

Management support of total quality by provision of appropriate resources and assistance

- Documented evidence of how training is funded/facilitated and is reviewed for effectiveness and efficiency
- Total quality structure includes guidelines on frequency of team meetings, the provision of suitable meetings facilities and allocation of funding for improvement activity
- Project management teams define priorities in improvement activity
- Funding of study tour to United States of America.

Management involvement with customers and suppliers

- Supplier assessment every six months as part of vendor-rating system
- Management meet with customers on a regular basis. Reciprocal visits facilitate information exchange
- Management aware of both internal and external customer requirements. Mechanisms for gathering information include site visits, questionnaires, marketing database.

Management promotion of quality management outside the company

- Presentations at local schools
- Willingness to accept visitors, provide site tours, etc.
- Involvement with quality research unit at local university
- Publications in local media.

Coordinating structure

Undoubtedly, the development of quality awareness within the workforce raises expectations. The 'high' should be taken advantage of by using the enthusiasm of the employees to achieve simple yet important gains. It is important to have a quality or total quality infrastructure in place. The role of the total quality manager need not necessarily be filled by the existing quality manager/director. On some occasions two roles can be beneficial, especially where the quality manager is concerned with maintaining standards, or where the organization is large, decentralized or divisionalized.

Figure 7.4 illustrates a suggested coordinating structure for the transformation process. The company's quality council in this structure has a membership of the most senior personnel, including board members. One of these people must be given personal responsibility for the continuous improvement initiative. It is particularly beneficial if this person is the managing director. The *council's* responsibilities include:

- Overseeing the quality planning
- Establishing success criteria
- Monitoring the success
- Providing leadership
- Facilitating the rest of the quality structure.

The appointment of a dedicated *total quality manager* is essential. Their responsibility is to oversee the implementation phase of the programme, act in a supportive role, ensure that the workforce are trained and monitor performance. This is done by:

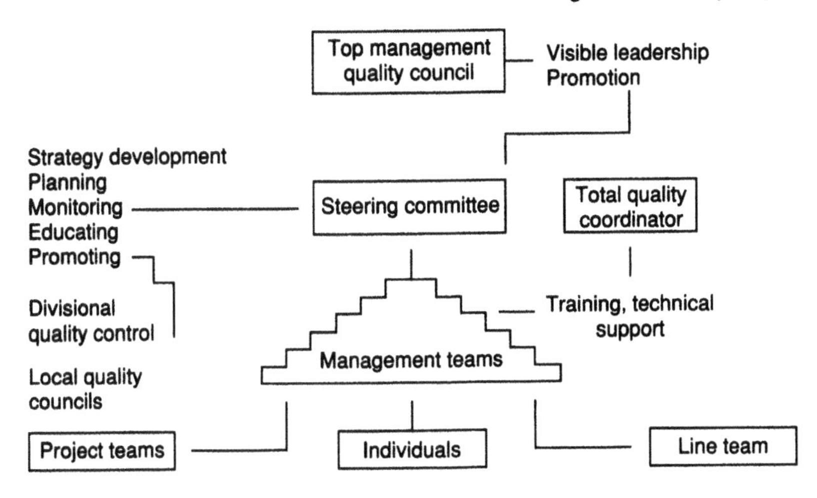

Figure 7.4 *Coordinating structure for quality*

- Acting as an adviser on all aspects of the total quality programme
- Developing with senior management the company's total quality strategy
- Managing all quality training
- Feeding back performance information to the company quality council.

Divisional quality councils only exist for large organizations where the resource is too big to be managed by the company quality council. Their membership includes senior divisional managers. Their role is to:

- Lead and control quality-improvement activities within their own division
- Define and publicize divisional quality objectives (in line with company strategy)
- Determine the priorities for quality-improvement activities
- Allocate human and financial resources to nominated projects.

The *local quality councils* exist to control quality operations on a day-to-day basis and one made up of local area managers. Their job is to:

- Steer and lead total quality functions within their own functions
- Provide support and guidance for total quality activities
- Determine local improvement priorities
- Select team leaders for projects
- Monitor team performance
- Ensure that any remedies are implemented
- Ensure that cost savings exceed outgoings.

Finally within the structure are the *quality improvement teams*. Without doubt, 'the expert' is the person who performs the job to the

approved standard, who is aware of the problems and of how to solv
them if we would only let them. Quality improvement teams ar
empowered workers. These teams are not limited to shopfloor person
nel. The managing director can sit on one of these teams if necessary
They tend to comprise four to six people who 'live' close to the probler
area. Their responsibility is to:

- Use the quality improvement tools and techniques to get to the roc cause of the problem
- Make recommendations to the local quality council
- Take corrective action as necessary
- Monitor the changed process.

Establishment of improvement projects

In the coordinating structure in Figure 7.4, the activities which fal under the responsibility of the line manager are presented at loca quality council level. A smaller organization, however, although i may have a steering committee comprising all managers, which under takes the work of a divisional quality council, will still require man agers to manage the quality-improvement teams. In that respect th responsibilities of the local area managers are still applicable.

Once senior management, in the form of a quality council or steering committee, have devised strategies for implementation, mid dle managers and line managers are responsible for the internal prior itization, coordination and monitoring of improvement projects. This i a carefully planned and directed effort to achieve a significant break through, resulting in a measurable improvement of a product, servic or process. The results may involve the solving of a problem, th reduction of costs or wasted material, reduced errors and rework less variation, etc.

As they are part of the important introduction stage the selection o improvement projects should be given careful consideration. They should have the potential for high visibility, cost savings and/or direc impact on customer satisfaction. To facilitate success initial efforts should focus on specific, tangible improvements to a clearly limitec and defined process. (For example, late delivery of a certain produc or elimination of product spillage in a bagging operation.)

The leader and members of the project teams are appointed by the middle/line managers. A prerequisite for team membership should be direct contact with the process under review. Team membership may move across divisionalized lines whenever this is applicable and neces sary for a representative team. Project teams may also draw from different levels of the organization when the scope of the project demands.

An important aspect of the improvement activities is that they be based on logical, analytical problem solving and the use of data. It is not sufficient for a project team to simply agree on some conclusion on the cause of a problem, for example. Relevant data should be accumulated to support any such conclusion. Data enables precise identification of the sources and causes of problems or of variation in the system, leading to the correction and prevention of problems at point of origin.

Project team management

Definition of problem

- Brainstorm list of problems
- Selection of problem
- Criteria for selection of problem – effect on costs, customer service, etc.

Initial analysis of problem

- Cause-and-effect analysis
- Determine most likely causes of problem
- Determine what data to be collected
- Data collection and analysis
- Data collection procedures (survey, questionnaire, audit, etc.)
- Analysis of data to demonstrate issues involved – this analysis can involve the use of Pareto Analysis or other tools (see Chapter 9).

Solution generation

- Possible solutions
- Illustrating criteria against which solutions were measured and a particular problem was chosen
- Describe proposed solution
- Outline advantages of solution
- Cost-benefit analysis.

Plan of action

- How is solution to be implemented?
- Expected completion date?
- Expected results?
- How solution will be evaluated?
- It is important to recognize that an emphasis on project and improvement activities does not constitute transformation. Projects are an important improvement tool, a useful team-building mechanism

and a worthwhile method of education. However, even a multitud
of project teams allied to one another in a mutually supportiv
manner will not transform an organization (see Figure 7.5).

Project title	Team leader	Start date	Estimated completion date	Project progress (stage reached)												
				Jan	Feb	Mar	April	May	June	July	Aug	Sept	Oct	Nov	Dec	Jar
						1	1	3								
						1	2	3								
						1	1	2								
						1	2	3								
						3	3	4								

Project monitoring stages

1 Select team
2 Define problem
3 Gather data
4 Analyse data
5 Generate solutions
6 Select best solution
7 Implementation plan
8 Implementation and recommendations
9 Set audit procedures
10 Completion form and report

Figure 7.5 *Project monitoring*

Coordinator support

The process of transformation involves the time, energy and commit
ment of resources throughout an organization. Someone must oversee
the logistical administrative and advisory processes involved in imple
menting quality. These operations encompass the establishment o
processes for the coordinators, technical assistance, training anc
review needed to support quality-improvement efforts.

- Help employees to assess the impact of various transformatior efforts, to determine what is effective and what is necessary; tha the right things are being done and are being done successfully
- Keep track of the progress of various improvement efforts. Asses needs. Coordinate any centralized training
- Coordinate the deployment of in-house technical resources. Provid for the continuous education of these resources
- Arrange for seminars and workshops for managers. Arrange fo managers and others to be presenters and instructors for variou workshops and seminars

- Provide technical assistance to the project teams and others engaged in improvement efforts. Assist in the establishment and education of teams of managers to oversee the project teams
- Provide technical support for the quality effort. Managers will perhaps need the advice of technical experts within the organization
- Provide orientation to new managers and other key participants
- Maintain a library of information resources and training materials
- Coordinate publicity for the transformation efforts: newsletters, professional journals, trade publications, local media, etc.

Such a coordinating function should begin modestly and evolve in an incremental, understated way. Initially, this function could be one person, a total quality manager, who assumes responsibility for all aspects of implementation, and who reports to the chief executive or managing director. Depending on the size of an organization, there may also be connecting, coordinating units developed. Ultimately, the coordinator function is an essential resource providing support services to managers. It is not responsible for the quality transformation.

Managing human resources

In the last 30 years, acquisitions, mergers and corporate restructuring have become a reality of business life. The increasingly competitive environment has resulted in managers seeking alternative means of dealing with the problems of redundancy, shorter working weeks, etc. The need has arisen for a more flexible response to the personnel needs of an organization. Increasingly, the traditional techniques of recruitment, selection and training are becoming focused not merely on achieving 'best practice', but also on 'best fit' to organization requirements. Participation in the process of forming strategic plans to deal with swiftly changing market conditions necessitates implementing effective human relation initiatives.

If the strengths of an organization's management determines its future there is a need to:

- Define, create and develop an organizational culture that attracts the personnel best suited to achieving corporate goals
- Invest in people (their recruitment, training and development).

Human resource planning, also known as manpower planning, can be important in monitoring costs while providing a productive workforce. No organization can rely on obtaining highly skilled personnel at short notice. Human resource planners are therefore essentially seeking to satisfy the business or economic objectives of the organization by finding the right types of skill, at the right quality, in the right quantities, in the right places at the right time. To achieve this, human

resource strategy, as already indicated, must be linked to the dail
activities of the organization and to business planning.

Consequently, human resource planning is no longer exclusive to a
small number of planning managers with a base on which to buil
plans for recruitment training, etc. Given that the human resourc
planning process involves:

- An analysis of the existing human resources supply (i.e. make-up o
  the current labour supply by estimating future supply both internall
  and externally, and
- Determining current labour resources by conducting a manpowe
  inventory by department, function, occupation/job title and leve
  of skill or qualification

managers from all levels and functions can contribute not only to
providing a better basis for planning employee development in orde
to make optimum use of employees but also to improving the overal
business planning process.

Recruitment and selection

The overall aim of the recruitment and selection process in an organi
zation is to obtain the quantity and quality of employees required by
the manpower plan. This process has three main stages:

- Definition of requirements, including the preparation of job descrip
  tions and specifications
- Attraction of potential employees, including the evaluation and use
  of methods of identifying sources of applicants from both inside and
  outside the organization
- Selection of candidates.

For the recruitment and selection process to be efficient a systematic
approach is required:

- Detailed manpower planning
- Job analysis
- Identification of vacancies
- Evaluation of the sources of labour
- Preparation and publication of information
- Processing applications
- Notifying applicants of results of the selection process
- Preparing employment contracts, induction planning programmes
  etc.

ob analysis

According to the British Standards Institution, job analysis is 'the determination of the essential characteristics of a job', i.e. the process of examining a 'job' to identify its component parts and the circumstances in which it is performed. This appraisal may be used for vocational guidance, personnel selection, training or equipment design. The product of the analysis is usually a *job specification* – a detailed statement of the activities (mental and physical) involved in the job, and other relevant factors in the social and physical environment.

Job analysis is a management problem-solving tool: e.g. to help make rates of pay more equitable, to direct reorganization for increased efficiency, to recruit more productive staff, etc. Job analysis, and the job specification resulting from it, may be used by a manager:

1 In recruitment and selection – i.e. for a detailed description of the vacant job
2 For appraisal – i.e. to assess how well an employee has fulfilled the requirements of the job
3 In devising training programmes – i.e. to assess the knowledge and skills necessary in a job
4 In establishing rates of pay – this will be discussed later in connection with job evaluation
5 In eliminating risks – identifying hazards in the job
6 In reorganization of the organizational structure – i.e. by reappraising the purpose and necessity of jobs and their relationship to each other.

The fact that a job analysis is being carried out may cause some concern among employees: fear of standards being raised, rates cut, redundancy, etc. The job analyst will need to gain their confidence by:

1 Communicating: explaining the process, methods and purpose of the appraisal
2 Being thorough and competent in carrying out the analysis
3 Respecting the work flow of the department, which should not be disrupted
4 Giving feedback on the results of the appraisal and the achievement of its objectives. If staff are asked to cooperate in developing a framework for office training and then 'never hear anything', they are likely to be less responsive on a later occasion.

Information which should be elicited from a job analysis is both task-oriented information and worker-oriented information, including:

1 Initial requirements of the employee: aptitudes, qualifications, experience, training, required, etc.

2 Duties and responsibilities: physical aspects; mental effort; routin
or requiring initiative; difficult and/or disagreeable features; cor
sequences of failure; responsibilities for staff, materials, equipmer
or cash, etc.
3 Environment and conditions: physical surroundings, with particu
lar features – e.g. temperature or noise; hazards; remuneratio
other conditions such as hours, shifts, benefits, holidays; caree
prospects; provision of employee services – canteens, protectiv
clothing, etc.
4 Social factors: size of the department; teamwork or isolation; sort o
people dealt with – senior management, the public, etc.; amount o
supervision; job status.

Job analysis is also used to determine job-performance standards
These standards serve two functions. First, they become objectives o
targets for employee efforts. Second, standards are criteria agains
which job success is measured and enable a control system to b
developed which can evaluate job performance (see Figure 7.6). Job
performance standards are developed from the job analysis informa
tion and then actual employee performance is measured. When mea
sured performance deviates from the job standard, corrective action i
taken.

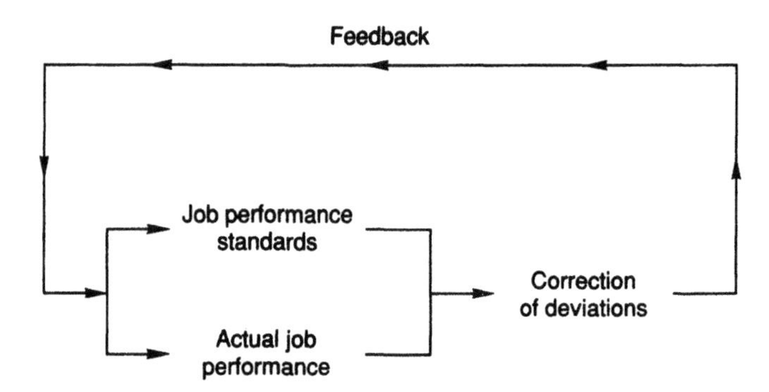

Figure 7.6 *Control system*

The 'scientific management' theory advocated organization struc
tures based upon conventional organization charts, functional specia
lizations and a hierarchy of positions with set roles and defined areas o
responsibility. Early management theory depended on the writings o
an American engineer called Frederick W. Taylor,[5] who was concerne
with efficiency and tried to apply scientific and engineering principle
to the work of people. His views encompassed the notion of 'time anc
motion' and stressed such factors as planning work and allocatin
responsibility, as well as part of the remit of management
Consequently, traditional approaches to job analysis assumed tha

jobs of work should be structured and stable. However, research has demonstrated that there is no one ideal type of organization structure. Furthermore, trends have been identified which are bringing about change and a need for coherent human resource strategies, i.e.

- Changing markets
- Technological development
- Evolving social norms.

The general implication is that a more flexible approach is required and, in relation to job analysis, a more 'organic' approach is needed in such dynamic situations. Traditional job descriptions are appropriate to stable circumstances as individuals and jobs have to be matched. However, job descriptions should not simply be used in order to achieve results. They can be employed as a method of education to help employees to understand how and why an organization works. They must emphasize not only the purpose of the job but also the importance of initiative and the need to respond to change.

Recruitment

Recruitment involves searching for and obtaining qualified job candidates in sufficient numbers that the organization can select the most appropriate person to fill the job needs. It is defined as the set of activities used to obtain a pool of qualified job applicants. There are several purposes of recruitment:

- To determine the present and future recruitment needs of the organization in conjunction with human resource planning and job analysis activities
- To help increase the success rate of the selection process by reducing the number of underqualified job applicants
- To meet the organizational responsibility for legal and social obligations regarding the composition of its workforce.

Although organizations may use both external and internal recruitment sources, they may not always obtain the desired number of applicants or retain the most valued employees. Organizations are increasingly realizing that it is essential to engage in proactive, strategic human resource planning to better manage the following areas:

- *Career development*: Exposing minority employees with high potential to the same key development jobs that have led traditionally to senior positions for white, male counterparts
- *Diversity training for managers*: Addressing stereotypes and cultural differences that interface with the full participation of all employees in the workplace

- *Diversity training for employees*: Helping employees to understand th corporate culture requirements for success in the company an career choices available to them
- *Upward mobility*: Breaking the 'invisible' (or) 'glass ceiling' an increasing the number of minorities in upper management throug mentors and executive appointment
- *Diverse input and feedback*: Asking employees themselves what the need as against asking managers what they think employees need

Selection

Selection and placement procedures provide the essence of an organi zation – its human resources. Selecting employees likely to perforn well may result in substantial productivity gains and cost savings Serving an organization's needs and being effective with selection and placement mean attaining several purposes:

- To contribute to the bottom line through efficient and effective pro duction
- To ensure that an organization's financial investment in employee pays off
- To evaluate, hire and place job applicants in the best interests of bot the organization and the individual
- To make decisions with consideration of the uniqueness and th individual, the job, the organization and the environment – perhap even to the extent of adapting the job or organization to the indivi dual.

Selection techniques

The selection interview is probably the most widely used assessmen technique, despite the fact that it has come into disrepute among recen systematic researchers. The research of Carlson[6] shows that the typica unstructured interview tends to be unreliable and therefore invalid and that assessments generated depend more upon the attitudes o the interviewer than on the characteristics of the interviewee. Thes systematic studies illustrate the following negative aspects of the typi cal selection interview.

- Interviewers have their own personal stereotypes for what constitu tes a good potential employee which may bear no necessary relation ship to a valid profile of what this is
- Interviewers tend to make a decision in the first 5–10 minutes of th interview and then spend the remainder of time seeking informatio to substantiate this initial impression

- Applicants' negative information is not equally weighted with positive aspects and therefore has greater impact on an interviewer's judgement
- An interviewer's judgements are not solely determined by the applicant's characteristics, but are also influenced by the relative strength or weakness of immediately preceding interviews and the presence or absence of numbers of applicants who must be hired
- Interviewing experience does not result in interviewers' judgements becoming more accurate and valid unless they receive systematic feedback regarding the strengths and weaknesses of those people they have interviewed.

Most organizations, however, are aware that interviews have a variety of positive characteristics, including the opportunity to provide information to the applicant regarding the position in particular and the organization in general. It may be used as a public relations device by the organization to help to project a positive image. Probably the greatest aspect of the interview, however, is its flexibility, which provides the organization with an opportunity to fill in the gaps of incomplete information that may be necessary for a decision. Interviews permit interviewers to make assessments of their compatibility with the applicant, and this may be especially valuable if the interviewer will be the interviewee's immediate superior. Finally, the interview may in some cases be the only appropriate form of assessment available. In most professions and senior positions in an organization, the most desirable potential candidates may be unwilling to subject themselves to any other form of assessment.

The interview as a selection device is fraught with problems and pitfalls, many of which can be minimized by training interviewers, standardizing the structure of the interview, keeping systematic written records and using multiple interviewers whenever possible.

Exercise 7.1

Have you ever been involved in staff interviews? What strategy and structure was followed? What training did you receive? How could the interview process have been made more effective?

Tests

Tests as a selection technique have the advantages of being quick, inexpensive and objective. These advantages, combined with the possibility of designing tests which predict job performance accurately, have resulted in the development of a large number of tests for use in personnel selection. The types of tests used in organizations for selection can be broadly classified into two categories. The classification is based upon the view that a person's performance in an organization is determined jointly by a combination of the person's motivation and ability. Ability testing is useful in predicting who will subsequently be successful on the job. Motivation testing measures a person's pattern of interests and/or aspects of a person's personality.

Tests, when carefully constructed and validated, can aid in the assessment process. Carefully designed ability tests can frequently improve the quality of selection decisions. Motivation tests, while helpful for some types of jobs, are less generally valid. The capacity of any test to predict subsequent performance must be systematically evaluated by the organization. In addition, organizations must determine whether a test adds anything to their ability to predict success over and above that provided by alternative devices.

Biographical information banks (BIBs)

These are forms often used for assessing applications for most jobs below senior management levels, which request information regarding basic demographic information, educational history, work experience, job-related training, salary, etc. BIBs are impressive predictors of success on the job, and consistently outperform all the other methods of assessing applicants. They are a potentially effective and efficient assessment tool and, when carefully developed, properly weighted and systematically validated, BIBs demonstrate an impressive ability to predict job success. Their validity is particularly impressive in view of their low cost and relative ease of construction and validation.

The most commonly used methods of assessing job applicants are interviews, tests and BIBs. Interviews, though used almost universally suffer from a variety of bias and shortcomings which tend to make them poor predictors of job success. However, techniques are available for making interviews more systematic and useful as an assessment technique. Ability tests tend to be related to job performance, while, on the other hand, personality tests are much less frequently predictive of subsequent success on the job. BIBs, when carefully developed, can serve as strong predictors of job performance and are extremely cost effective.

Cost benefit assessment

Organisations need to determine whether the benefits of selection and placement decisions outweigh the costs. Benefits of good selection and placement decisions include higher sales and profits as well as higher quality of the product or service. The actual costs incurred in hiring applicants include:

- Recruitment and assessment costs – salaries of personnel staff
- Costs of personnel training – costs associated with the pre-assessment, implementation and evaluation phases of training
- Induction and orientation costs – administrative costs of adding employees to the payroll; salaries of new employees
- Training costs – salaries of training and development staff; salary of new employees during training.

Selection and placement decisions can be assessed by measuring employees' satisfaction with work, the extent to which they feel their skills and abilities are being used and their level of involvement with the job and the organization. As employee satisfaction, involvement and skill levels can change, human resource managers must continually monitor these activities. Regular, periodic organizational surveys are one method.

Education and training

Education must be a pervasive effort in the change process. An essential ingredient of the transformation is the development and expansion of training and education programmes. Specific training approaches will be addressed in Chapter 9. The following are some of the types of training needed to support the quality initiative.

- *Technical training related to specific job skills.* Everyone should have a mastery of the technical skills needed to do their job. All employees with an identical job should do it consistently, eliminating variation from one worker to another
- *Systems orientation for all individuals and groups.* All should understand how their jobs fit into the system, who their internal suppliers and customers are and how their work affects the final product or service delivered to the outside customer and user
- *New technical and maintenance skills.* Technical knowledge and skills previously reserved for technicians (for example, engineers and maintenance personnel) should be gradually transferred to operators. Technicians should be viewed as instructors for the hourly workers. In turn, the knowledge and skills of the technicians should be upgraded. The goal is to evaluate everyone's level of technical competence
- *Basic orientation to quality.* This includes presentations on such topics as the history of the quality movement, the essentials of quality and transformation, your organization's approach to transformation and your plan. These should be taught to everyone at an early stage of the transformation effort

- *Technical advisor training*. Early in the implementation of quality, an organization should begin developing an internal network of personnel who are capable of providing consultation and technical assistance to those engaged in improvement efforts. These individuals know the basic tools of the scientific approach, the skills of project planning and management and the basics of team development and meeting management. The technical advisors also know how to teach these skills to others
- *Basic improvement skills*. Gradually, everyone in the organization should learn these skills: How to plan and manage an improvement project. How to work in groups. How to plan change. The basic scientific tools. How to gather data to determine the sources of problems and variations.

Appraisal and development

Performance appraisal is defined as measuring, evaluating and influencing an employee's job-related attributes, behaviours and level of absenteeism to discover at what level the employee is presently performing on the job and in the organization. It is a formalized, legitimized process of observation and judgement. The typical performance appraisal system consists of the following elements:

1 *Standard of Measurement*: In order to avoid arbitrary evaluations, a manager seeks to establish a measurement standard against which he or she judges people's performance.
  - In the production or direct service areas, operators and supervisors are judged on some unit of output over some unit of time: tons per day, calls per week, etc.
  - Managers and professionals are involved in less routine activities and therefore have standards tailored to their work: completion of this project, successfully solving that problem, etc.

2 *A method for establishing the standard*: There are many ways to establish a standard. They can be based on:
  - Government regulations or customer specifications
  - Budgeted projections (sellable tons of produce *x* per month)
  - Past performance
  - National or industry standards
  - Machine capabilities specified by the equipment manufacturer
  - Negotiations and mutual agreements between supervisor and subordinate. This is more frequent among managers and professionals.

3 *A period of performance*: A performance-evaluation system specifies a period of time over which the accomplishments are to take place before the activities are reviewed and evaluated. For managers and

professionals, the period is normally one year, sometimes six months.

4 *A performance interview*: At the end of the time period there are usually one or more meetings at which the person's performance is discussed. How did he or she do during this period of time relative to the established standards of performance?
5 *The rating*: Finally, there is some form of rating which varies widely from company to company. Sometimes there are only a few categories ('Superior performance, Acceptable performance, Below standard performance'). Some ratings are on a point scale ('You rate 6 on a 10-point scale'). Often there are scales attached to categories of behaviour ('Courtesy: 1 2 3 4 5'). Sometimes the rating is in the form of a narrative that describes the employee's level of performance and makes recommendations for improvement. Some ratings are done verbally, with nothing in writing – no 'report card'.

Performance evaluation has its own internal logic. A manager is aware of what his or her division must accomplish, what goals and objectives have been determined for the division as its contribution to the progress of the whole enterprise. The manager, therefore, makes sure that the sum of all the standards and objectives of employees equals or exceeds the performance expected of the entire division.

Performance appraisal can have an impact on productivity and competitive strategy of the organization. It serves as a contract between organization and employee, a contract which in turn acts as a control and evaluation system better enabling performance appraisal to serve many purposes:

- *Management development*: It provides a framework for future employee development by identifying and preparing individuals for increased responsibilities
- *Performance measurement*: It establishes the relative value of an individual's contribution to the company and helps to evaluate individual accomplishments
- *Performance improvement*: It encourages continued successful performance and strengthens individual weaknesses to make employees more effective and productive so that organizations can successfully implement their strategies, such as quality enhancement
- *Compensation*: It helps to determine appropriate pay for performance and equitable salary and bonus incentives based on merit or results
- *Identification of potential*: It identifies candidates for promotion
- *Feedback*: It outlines what is expected from employees against actual performance levels
- *Human resource planning*: It audits management talent to evaluate the present supply of human resources for replacement planning

- *Communications*: It provides a format for dialogue between superio and subordinate and improves understanding of personal goals an concerns. This can also have the effect of increasing the trust betwee the rater and ratee.

However, Dr Deming, and subsequently others, argue against th 'apparent logic' of performance appraisal, claiming that it impedes th transition to a stable total quality environment.

1 *Any employee's work, including the work of managers, is tied to man systems and processes. BUT performance evaluations focus on individuals as if those individuals could be appraised apart from the systems in whic they work.* Performance evaluation does not make sense if individuals or groups are held responsible for events, behaviours, circumstances and outcomes over which they have no control. Yet an individual's performance almost invariably depends on man external factors. Managers depend on the state of the economy salespeople depend on the economy plus the quality of the product; line workers depend on the state of the machinery; word processing operators depend on the state of someone's penmanship and on and on. We all inherit elements of our work from those wh precede us in the process. On what then, should I be evaluated? Or the value I add to the work before I pass it on? If so, how will the success of my contribution be differentiated from all the contributions – positive and negative – that precede and follow me? The fac is, those who feel they can make such differentiations are ofter simply guessing.
2 *Most work is the product of a group of people. However, a process o evaluating an individual requires an assumption that the individual i working alone.* Almost nothing is accomplished by an individua operating alone. Most work is obviously a collective effort. Ever workers who seem quite independent depend on others for ideas stimulation, feedback, moral support and administrative service When someone is credited with a success, he or she is individually honoured for what was, most likely, the work of many. And it is no only praise that is distributed unevenly. Just as success is a grou effort, so also is failure. Having a system where it is individuals who are rewarded or recognized will force workers to choose between individual reward and recognition or the teamwork Such a choice will seldom reinforce teamwork. The choice is naturally divisive. There is an abundance of evidence that teams perform most activities more effectively than individuals engaged in the same task. The evaluation of individual performance undermines this teamwork.

Deming presents a view as logical against appraisal as those who support its use. Is there an alternative in total quality terms? If performance appraisal is to:

- Provide feedback to the employee
- Provide a basis for a salary review
- Help to identify candidates for promotion
- Give direction to an employee's work
- Provide an opportunity to give recognition, direction and feedback to an employee
- Identify an employee's needs for training, education and career development
- Provide a forum for communication

what does management have to consider in order to move beyond the notion that traditional appraisal vehicles are unable, in total quality terms, to assess individual performance accurately?

Activity	*Total quality appraisal*
1 Provide feedback to employees	Feedback from all involved in employee's work Identify major processes in which employee is involved Identify major work group(s) to which employee belongs Develop list of major feedback sources for each employee: • Peer feedback • Key customer feedback • Key supplier feedback Sole purpose of feedback is improvement
2 Identify candidates for promotion	Other possible methods: • Special assignments which contain elements of promotion job, to test level of ability. Monitor as per methods identified in providing feedback • Assessment centres – specially designed activities to assess candidates using skill needed in new position under similar conditions as will be found in new job. • Involve 'customers' of those who will be team members,

	customer, etc. of promotee. They can help to develop work criteria, selection methods, etc.
3 Give direction to an employee's work	Develop and communicate company mission and policy statements in order to define purpose and direction of organization which can guide everyone's work. • Managers should plan with their employees 'how' to achieve goals (i.e. in the next project, for the next two years the next five years, etc.) • Frequent *communication* between employee and supervisors facilitates better understanding of actual progress and potential problems.
4 Identify need for training, education, career development	Define job and its requirements which will assist experts in designing methods to determine capabilities for the necessary skills. Use casual communication to identify training opportunities.
5 Provide forum for communication	Possible communication media: • Meetings • Focus groups/feedback groups. *Ad hoc* meetings with specific purpose. Provide opportunity for collecting information from employees • Management by walking around (MBWA) 1 Observe activities 2 Ask employees to talk about tasks 3 Determine problems – what prevents better work?

Since appraisal really begins before people join teams, it should be oriented towards the employee achieving personal development and self-fulfilment. The appraisal process should continue until an employee leaves the organization and should include:

- Selecting individuals who have a natural ability to work within teams and subsequently building teams in which each member understands how their abilities contribute to the team's success. Psychometrics of ability and personality (eg Belbin[7]) may help in this respect
- A formal interview at least once a year, allowing all employees to gain a greater understanding of their aims, strengths and weaknesses
- Assisting all employees to develop skills and advance their careers
- Developing within leaders the capability to bring a team of people together who will strive to achieve common team objectives.

It would be ideal if appraisal took place every day. However, counselling with staff on a daily basis is not always realistic and indeed, focusing attention on one problem at a time may not take into consideration the broader needs of the employee.

Objectives of total quality appraisal

- Gain a common understanding of each individual's job. This may involve contribution to more than one process – people are often in more than one team at a time
- Gain insight into personal values and ways in which the job may be developed to increase personal satisfaction
- Gain insight into individual skills, strengths and weaknesses
- Begin communication between manager/supervisor and employee about a personal development plan.

It is important in an appraisal interview, be it annual or biannual, that both employee and manager prepare and present their view of the job and associated skills separately. The interview can then become a discussion forum, the aim being to bring both parties to a consensus on how to improve.

The motivation to work is a key factor in achieving high performance. Explanations of what causes motivation often contradict, and while the main schools of thought (Process, Content and Behaviourist) all offer differing insights into the needs of the individual there are other factors which may inhibit performance such as:

- The task may be physically or mentally impossible for the individual
- The physical resources may not be available
- The individual may not have the ability to do the task.

A checklist for the diagnosis of performance problems has been developed by Wright and Taylor[8] as shown on pages 156 and 157.

This checklist incorporates the good practices identified by the exan ination of motivation theories.

Checklist for improving work performance

- What is the problem in behaviourial terms? What precisely is th individual doing or not doing which is adversely influencing his c her performance?
- Is the problem really serious enough to spend time and effort on?
- What reasons might there be for the performance problem? (Se Column 1 below.)
- What actions might be taken to improve the situation? (See Column below.)

Column 1 *Possible reasons for performance*	*Column 2* *Possible solutions*
Goal Clarity Is the person fully aware of the job requirements?	Give guidance concerning expected goals and standards. Set targets.
Ability Does the person have the capacity to do the job well?	Provide formal training, on-the job coaching, practice, secondment, etc.
Task difficulty Does the person find the task too demanding?	Simplify task, reduce workload decrease time pressures, etc.
Intrinsic motivation Does the person find the task rewarding in itself?	Redesign job to match job holder's needs.
Extrinsic motivation Is good performance rewarded by others?	Arrange positive consequence for good performance and zero or negative consequences for poor performance.
Feedback Does the person receive adequate feedback about his/her performance?	Provide or arrange feedback.
Resources Does the person have adequate resources for satisfactory task performance?	Provide staff, equipment, raw materials as appropriate.

Working conditions

Do working conditions, physical or social, interfere with performance?	Improve light, noise, heat layout, remove distractions, etc. as appropriate.

- Do you have sufficient information to select the most appropriate solution(s)? If not, collect the information required, e.g. consult records, observe work behaviour, talk to person concerned
- Select most appropriate solution(s)
- Is the solution worthwhile in cost-benefit terms? If so, implement it. If not, work through the checklist again, or relocate the individual, or reorganize the department/organization, or live with the problem
- Could you have handled the problem better? If so, review your own performance. If not, and the problem is solved, reward yourself and tackle the next problem.

Reward and recognition

The reward system is one of the most emotive issues to be confronted in the course of organizational change. A reward and recognition programme is established in order to encourage employees to work to achieve company or departmental goals. Employees require a clear understanding of goals in order to achieve them. An important preparatory step in developing a reward and recognition programme is the establishment of organizational mission, strategy and associated goals (see above).

The clear definition of goals, at organizational, departmental and individual level, will help to develop a programme that is consistent with organizational goals. Once organizational goals have been developed each department should determine how to achieve the objectives relevant to its activities and personnel. Measuring performance before the programme begins provides a basis for measuring the success of the programme. The measuring of performance against set goals and objectives tends to a clear understanding of current performance and the setting of realistic targets.

In order to enable employees to reach or succeed targets set, it is important that they are supplied with the relevant tools in a cooperative environment which is conclusive to teamwork. However, in order to gain benefit from a reward and recognition programme, it is important to review the current reward system in order to determine if it works against the quality ethos. If an effective pay system is not in place, employees may resent a reward and recognition programme. Employees who do not believe that pay, determined by a merit system, is based on performance may not feel motivated by the reward and recognition incentive. If your reward system is based purely on quality, then it may provide an incentive in that basic or poor perfor-mers receive the same as good performers.

The key principles of a reward system are:

- It should reinforce desired behaviour and achievement
- It should discourage undesired behaviour
- It should allow individuals the opportunity to influence the para-meters that make up performance-based aspects of the system.[9]

Reward criteria

- Clearly stated objectives
- Quantitative reward criteria – measurement of achievement agains objectives promotes fairness
- Achievable rewards – too high standards can discourage.

Recognition

The key principles of recognition are:

- Determine what a recognizable achievement is and develop groun rules
- The recognition should be relevant to the achievement and not statu within the organization (i.e. applicable for *all* levels)
- Senior management presence at presentations
- Communication of achievement to others.

The actual format of recognition can vary and the 'culture' of a organization must be considered when choosing symbols of recog-nition. Depending on the organization, recognition can fall into th following categories:

- A verbal 'thank you'
- Pens, calculators, gifts, etc.
- Presentations to management
- Newsletters
- Notice boards
- Employee of the month awards
- Visits to other companies
- Articles in the local press
- Social functions with senior managers for high achievers an partners.

Conclusions

In order to transform quality into a strategic business plannin and management dimension the quality function should be th organization's focal point and gauge of a customer's quality expectations and subsequent satisfaction. Normally, organization systems are viewed in terms of internal dynamics as they apply to design, production, marketing, etc. A more all-encompassing system has developed, integrating the interests and needs of both customers and suppliers. Consequently, management at all levels has to develop a more in-depth understanding of the organization as a system.

These considerations lead to an evolving profile of the manager responsible for quality. Each function, as it becomes responsible for controlling and improving its own processes, products and services, requires guidance and direction. Their efforts need to be facilitated on a daily basis. The quality manager:

- Must clearly understand the organization, its processes and interfaces. This manager must be conversant with the key functional languages used in the business, and comfortable operating at many organizational levels
- Should be treated as, perceived as, and be a valuable resource to the management team
- Must fully understand, and be an effective advocate and teacher of, total quality. This implies a willingness to relinquish many traditional responsibilities
- Should be people-oriented and patient, able to balance organizational realities with competitive necessities
- Should be able to shift emphasis from problem solving to motivating and teaching others how to solve their own problems
- Must be flexible, and become an effective agent of change in a rapidly changing world.

Assessment activity

The objective of this assessment activity is to assist in the development of an adequate infrastructure for quality transformation. The issues addressed in this chapter provide the guidelines for the approach.

1 Do any of the elements of a total quality infrastructure exist within your company?
   - Communication channels
   - Worker–manager relations
   - Systems and procedures
   - Standards of performance
   - Staff development
   - Team formation
   - Commitment
   - Training and education
   - Focus on customers

   Are they adequate?

2 Devise an organizational chart for your company indicating th positions of various teams/functions (i.e. quality councils, improve ment teams, etc.)
3 What will the roles of various levels of management be? Is th overall style of management conservative or progressive? Wha evidence could you use of indicators to confirm this belief? How can the managers of marketing, purchasing, personnel, engineering etc. lead the quality efforts in their functional areas? How is resis tance to change most likely to manifest itself? How can such resis tance be counteracted?

Summary points

- The quality organization must be viewed as a system of interrelate activities.
- Employees are not viewed as 'points of control' but as willing parti cipants in the ongoing process of improving organization perfor mance.
- The establishment of a coordinated structure for the transformatio process is essential.
- The process of transformation requires senior management commit ment.
- The skills and capabilities of individuals must be preserved throug recruitment, training appraisal and reward management.

References

1 Margulies, N. (1973), *Organisational Change, Techniques and Application*, Illinois: Scott Foresman.
2 Bower, J.L. (1972), *Managing the Resource Allocation Process*, Illinois: RD Irwin.
3 Deming, W. E. (1980) *Out of the Crisis* MIT, Cambridge, Mass: Centre for Advance Engineering Study.
4 Kubler-Ross, E. (1986), 'On Death and Dying,' OD, *Practitioner*, **18**, December.
5 Taylor, Frederick W. (1911), *Principles of Scientific Management*, New York: Harper & Bros.
6 Carlson, R.E. (1967), 'Selection interview decisions: the effect of interviewer experienc and applicant sample on interviewer decisions', *Personnel Psychology*, **43**, 259–280.
7 Belbin, R.M. (1981), *Management Teams, Why They Succeed or Fail*, Oxford: Butterworth Heinemann.
8 Wright, Peter and Taylor, David (1984), *Improving Leadership Performance – A Practica Approach to Leadership*, Englewood Cliffs, NJ: Prentice-Hall International.
9 Hillman, P. (1992), 'It's simple, but not easy', *Managing Service Quality*, September.

8 Teamwork for quality

Introduction

For many years the search for ways to improve management performance has focused almost entirely on the individual. Organizations have been preoccupied with the qualifications and experience of individuals even though no individual can combine all the essential qualities needed for a particular task. However, a team of individuals certainly can, and for this reason, and because organizations now recognize the need to involve all personnel in continuous improvement activities, organizations are beginning to emphasize teams as a mechanism for improving performance.

A team is a group of two or more people who work together towards the achievement of a common goal. Teams can offer the blend of personal attributes necessary for success and, more importantly, enable all employees to contribute to the future success of the organization. The adoption of a team-based approach offers many other advantages:

- Synergy, where the output of the team is greater than each of the separate inputs or the sum total of the inputs
- Teamwork is an effective mechanism for promoting change as people are more likely to take ownership of a solution or change that they have been involved in
- Improved communications within and between departments, building trust, and generating a spirit of collaboration rather than confrontation in the organization.

Those who have been associated with teams will recognize that certain conditions have to be met if they are to be successful. This chapter will focus on the key issues and how they can be addressed. Its objectives are:

- To review the most common types of quality team
- To discuss the factors crucial to team success
- To understand the ground rules for team selection and development
- To identify the characteristics of successful teams.

Types of team

Quality improvement teams often take the form of either process teams or task teams. Process teams work to improve a process and include

people who work within or support a process. The team exists as lon
as the process exists, even though team members may change as th
improvement effort changes. Selected individuals, with specialis
knowledge about a part of the process, may be included as part-tim
team members and their attendance may vary, depending on th
team's need for their services.

Task teams perform a specific task or solve a specific problem an
can take many forms. The most popular forms are the project team an
quality circle. Project teams usually comprise members from differen
functional areas and disciplines who meet on a part-time, but regula
basis to complete a task often assigned by management. On completio
of the task the team normally disbands. A quality circle usually consist
of members from a single functional area who meet on a regular basi
to solve everyday work problems chosen by themselves.

Task team membership is usually on a voluntary basis and the team
themselves normally meet during working hours. Their objective i
often to identify, analyse and solve work-related problems. In man
organizations their involvement ends with a presentation of their pro
posals to the management team but they should also be given respon
sibility for implementing their solutions. It is only through the activ
involvement of those who will have to work with the solution that i
can be implemented with minimum disruption and maximum effect.

The stages of team development

Teams pass through several phases during their development, display
ing common characteristics as they do so. Woodcock[1] has developed
simple model defining the four phases of development and the char
acteristics associated with them. Teams can use the model to under
stand where they are in the development process and where they wan
to be.

The undeveloped team

An undeveloped team is one that has devoted little time to considerin
how best it can operate and how the strengths of the team can b
capitalized upon. The way in which team members interact with on
another and the building of strong relationships are largely ignored
the team members operating as individuals rather than as a team. Th
basic ground rules for the team may be established and an initial tean
structure may emerge, although a shared understanding of what has t
be done will be lacking.

Such teams must be encouraged to share their views and experi
ences, and the team members must begin to communicate with on
another in an open and honest way. They should be encouraged t

consider each individual's strengths and weaknesses and to support one another if they are to develop into a mature and effective team.

The experimenting team

When a team begins to review its operating methods and to investigate new ways to improve its performance it can be described as having reached an experimenting stage. Team members will begin to examine interpersonal issues and their individual roles, and some members of the group may begin to resist the emerging team structure. The team members will attempt to retain their individuality and there will be a general lack of unity in the team.

Teams at this stage of their development should be encouraged to question team performance and to encourage each team member to participate in decision making and problem solving. The team should focus on conflict resolution through open and honest discussion in an effort to generate understanding and teamwork.

The consolidating team

During this phase the team members will begin to accept their role and to understand that of the others. Members will begin to listen to each other's point of view and to develop a spirit of cooperation. As team member relationships are built, trust will begin to grow and the team will begin to consider how they can develop a more methodical and agreed way of working. They will also begin to focus on their objectives, planning their activities and reviewing success.

Teams at this stage of their development should be encouraged to identify individual team-member weaknesses and how these can be compensated for by other members of the team. Regular reviews of performance and of potential improvements should also be initiated.

The mature team

Having improved their relationships and developed a more systematic approach, the team will begin to mature. They will become more flexible, internal politics will have been set aside and a high degree of trust will be evident. The team will be more focused on common objectives, will cooperate to achieve these objectives, will actively participate in decision making and will continually review the results being achieved. On an individual level, the team members will be more satisfied with their role and will find team membership more rewarding.

Characteristics of mature teams

As teams mature and pass through the four development phases, they begin to take on four important characteristics – a role structure, norms, cohesiveness, and informal leadership. These characteristics are important to team success and are discussed below.

Role structures

Each team member has a part to play, or role, in helping the group meet its objectives. Some people are leaders, some do the work and some interface with others outside the group. Furthermore, some people may be good at concentrating on getting the job done whereas others may provide the support the other team members require to function as a team. Too many of one type in a team means a lack of balance; too few roles and some tasks do not get done. In a small team one person may perform more than one role whereas in larger teams a full set of individual roles may be present.

In mature teams, each team member has a clear role to perform based on individual capability. The number of roles that the team member is expected to take on at any one time are also controlled. This ensures that expectations are clear, reasonable and consistent and that the individual's capabilities and limits are not exceeded.

Behavioural norms

A second major characteristic of teams is their norms or standards of behaviour. Norms define the boundaries between acceptable and unacceptable behaviour and can be specific to the team or work area. Norms can sometimes dictate role structures in that they prescribe roles for different team members (e.g. the junior member of the team performs the trivial tasks for the rest of the team). As newcomers join the team they gradually adopt the norms until eventually they conform to the team culture.

Cohesiveness

Cohesiveness is the extent to which team members are loyal and committed to the group. In a highly cohesive team the team members work well together, support and trust one another, and are generally effective in achieving their goals. In contrast, a team that lacks cohesion is poorly coordinated, its members may not fully support one another and may have difficulty meeting their objectives.

Five factors increase the level of cohesiveness in a team and five factors are known to reduce team cohesiveness:

Factors that Increase cohesiveness	*Factors that reduce cohesiveness*
Interteam competition	Team size
Personal attraction	Disagreement on goals
Favourable evaluation	Intra-team competition
Agreement on goals	Domination
Interaction	Unpleasant experiences

(Adapted from Griffin[2])

One of the strongest factors which increases cohesiveness is interteam rivalry, i.e. when two or more teams are in direct competition. Personal attraction seems to enhance cohesiveness and receiving a favourable evaluation by outsiders can increase it. Similarly, when there is agreement and acceptance of goals and frequent interaction among team members the team is likely to become more cohesive.

Size is an important factor in teams and is usually a matter of compromise. On the one hand, there is a need to widen the composition to bring in more knowledge and ability. On the other, there is a need to maximize involvement and individual effectiveness by keeping the team size small. Cohesiveness tends to decline as team size increases. It will also be lost if there is disagreement on what the goals of the team should be or if team members begin to compete among themselves, focusing on their own actions and behaviours rather than those of the team. Domination by one or more members of the group may also cause overall cohesiveness to decline as team members may begin to feel that they are not being given an opportunity to contribute. Finally, if the team has an unpleasant experience (e.g. they fail to achieve their initial objectives) they may become less cohesive.

Informal leadership

Although teams normally have a formal leader who has been appointed by the organization, the team members may also look to others for leadership. An informal leader is one who engages in leadership activities although this right has not been conferred by the organization.[2] Such an individual is likely to be good at getting the job done while satisfying the emotional needs of the team members. Therefore in situations where the formal leader can fulfil only one of these roles an informal leader may emerge to supplement the formal leader's functions. Such people can be an asset to the team where they work toward meeting the goals of the team. However, where this is not the case they can disrupt the team and lower performance levels.

Characteristics of effective teams

In attempting to identify the characteristics of effective teams one has t consider both productivity and member satisfaction as measures o effectiveness. The individual will judge the performance of the tean in terms of meeting personal expectations and needs such as friendshi and security, whereas team productivity is more likely to be judged i terms of goal achievement. As there is a correlation between membe productivity and satisfaction, organizations have begun to focus o both in an attempt to improve team effectiveness.

Larson and LaFasto[3] have identified eight properties of successfu teams.

A clear understanding of objectives

To develop a high-performance team there must be a clear understand ing of the problem to be solved and a belief that the results will b worth while. A clear focus on team goals and objectives and a sense o urgency must also be maintained. These objectives must be commu nicated in a way that will generate team commitment to achievin them. Politics and personal agendas must be excluded from the tean as these are the greatest threats to goal clarity and effective teamwork

Team structure

A key factor in the success of a team is the structure of the team itsel Consideration should be given to the overall objective of the tean before the team structure is formulated to ensure that individual an team efforts lead toward the desired goal.

When the team has been formed to resolve problems on an ongoin basis the most important feature of the team is trust. Each member o the team must have a high degree of integrity and be trustworthy s that good working relationships can develop. In addition, each tean member must be valued and treated with respect. A typical example o such a team is a quality circle.

Where the broad objective of the team emphasizes creativity then a necessary feature of the team is autonomy. Since creative teams need t be able to explore various alternatives, they should not feel restricte by the existing organizational structure or work in an atmosphere i which ideas are prematurely quashed. The Lockheed 'Skunk Works' i a typical example of such a team.

If a well-defined plan is being worked to there must be high tas clarity and clearly defined roles and responsibilities. Performance stan dards should be clear and there should be a high degree of responsive ness from team members. A surgical team is a typical example.

Once the team structure has been conceived the roles, responsibilities and accountabilities of each team member must be defined. The success of the team will depend on the collective performances of each individual, so each team member must be aware of what they will be held accountable for and how their performance will be measured.

Team member selection

It is imperative to select the right people if the team is to be successful. People are often chosen for the wrong reasons (e.g. because of their position or level of responsibility). Such considerations do not lead to successful teams. Instead, the most important factor is to select people who are best equipped to achieve the team's objective (i.e. those with relevant competencies). A competent team member is one who possesses the necessary technical skills and has the personal characteristics required to enable the team to meet its objectives.

Technical competencies are essential requirements for any team and relate to the knowledge and skills that the team members must have to meet their goals. It is essential to identify those technical skills required and the balance needed on the team. As would be expected, each team objective presents its own unique set of technical challenges and hence team technical requirements.

Personal competencies refer to the qualities and abilities necessary for the individual team members to work as a team. The types of individuals, their qualities as people, their specific talents, and their abilities to work together toward a common goal are critical determinants of team success. Good teams are able to identify all types of issues related to the team's objective, discuss them in a fact-based way, and resolve them via collaborative effort.

Belbin[4] has made a study of the optimum mix of characteristics in a team resulting in the identification of nine roles necessary for an effective team. These are:

- The coordinator
- The shaper
- The resource-investigator
- The monitor-evaluator
- The team worker
- The completer-finisher
- The plant
- The implementer
- The specialist.

The coordinator is the one who presides over the team and coordinates its efforts. Coordinators do not need to be creative but are disciplined, focused and balanced. They talk and listen well, and are good judges of people and of things. They rely on working through others.

The shaper may be highly strung, outgoing and dominant. Shapers are often task leaders and would, in the absence of the coordinator leap into that role. Their strength lies in their drive and passion for the task, but they can be oversensitive, irritable and impatient. However they are needed as the spur to action.

Resource-investigators are popular members of the team, extrovert sociable and relaxed. They bring new contacts, ideas and developments to the group and suit a liaison role. Such individuals are not original or drivers, and therefore need the team to pick up their contributions.

Monitor-evaluators are intelligent, but in an analytic rather than creative sense. Their contribution is the careful dissection of ideas and the ability to see the flaw in an argument. They may be less involved than the others but are necessary as a quality check. They tend to be dependable but can be tactless and cold.

The team worker holds the team together by being supportive to others, by listening, encouraging, harmonizing and understanding. Likeable and popular but uncompetitive, the team worker is the sort of person who remains unnoticed when there but who is missed when absent.

Without a completer-finisher the team might never meet its deadlines. These individuals check the details, worry about schedules and encourage others with their sense of urgency. Their relentless follow through is important but not always popular.

Unlike the shaper, the plant is introverted but is intellectually dominant. Plants are often the source of original ideas and proposals, being the most imaginative as well as the most intelligent members of the team. They can, however, be careless of details and may resent criticism. They also need to be drawn out or may switch off.

Implementers are the practical organizers who turn ideas into manageable tasks. Methodical, trustworthy and efficient, they are adept at administering. They tend to be unexcited by visions and can be unexciting themselves.

Specialists are dedicated individuals who pride themselves on acquiring technical skills and specialized knowledge. While they show great pride in their own subject, they usually lack interest in other people's. They have an indispensable part to play in some teams, for they provide the rare skill upon which the firm's service or product is based. As managers, they command support because they know more about their subject than anyone else and can usually be called upon to make decisions based on in-depth experience.

A spread of desirable personal attributes offering wide team-role coverage is important. This does not mean that the team must have nine separate role types, indeed one person may perform more than one role. What is important is that the membership offers a good spread in likely team roles with a wide range of team-role strengths on which to draw.

The personal competency requirements vary from team to team, although common characteristics can be identified. People who are intelligent, i.e. conceptual enough to see relationships and analytical enough to reduce problems to meaningful issues, do well in problem-resolution teams. People who understand 'the system' and can get things done, who possess a high degree of integrity and can maintain good human relations within the team are essential if trust is to be established and maintained.

Creative teams possess a range of personal characteristics that are somewhat unique. Not only do they require conceptual and analytic thinkers, but the team members must also be independent thinkers, capable of looking beyond what is currently accepted for less traditional solutions to problems. Self-starters who take a personal interest in the team's objective and who are willing to make personal sacrifices in order to meet the objective are also required. Finally, members of creative teams usually possess a high degree of confidence and tenacity when dealing with difficult problems with no immediate solutions.

Tactical teams, which focus on the operational execution of a task, require team members who are dependable, consistent and precise in their approach. Such teams require individuals with commitment, who are highly action oriented and who possess a sense of urgency.

Commitment

Unified commitment is often the most clearly missing feature of ineffective teams and refers to the sense of loyalty, enthusiasm, dedication and identification with the team. Anundsen[5] concludes from her analysis of teamwork that teams do not excel without serious individual investment of time and energy. Team success comes from everyone working together with a genuine dedication to the goal and a willingness to expend energy to achieve it.

Commitment can be fostered within a team in a number of ways, such as:

- Encouraging team members to set high expectations for each other
- Encouraging and expecting everyone to contribute to their full capability
- Ensuring that individual objectives are not pursued at the expense of the team goals
- Ensuring that individual performance problems are resolved
- Ensuring that team members are recognized for their contributions to team success.

Collaboration

Teamwork is an important factor in determining a team's success an can result in the team achieving more than could be expected from knowledge of the members' skills and abilities. It can be developed b building strong relationships between the team members and the tea leader. A climate of honesty and trust needs to be developed, allowin the team members to focus on the problem instead of diverting the attention to other matters. Individuals must be prepared to be ope and honest and to confront issues as they arise if they are to meet the team goals. This spirit of cooperation and trust encourages each tear member to help the others to overcome obstacles and to compensate fo one another, resulting in a superior team performance.

Trust and collaboration come about as a result of involvement an autonomy. The team members must be involved in planning an strategy development so that they know what is expected and how i is to be achieved. In addition, they should be given the autonomy t implement the strategy without the need for constant referral to other outside the team. Respecting people's rights as individuals and tean members will also encourage collaboration and teamwork.

High standards

Effective teams tend to set high standards and then strive continually t improve upon them. Such teams often expend a high degree of indivi dual effort in meeting these standards and indeed the members ofte set high personal standards. The team leader's ability to inspire th team plays an essential part in achieving high standards. In additior the pressure exerted by team performance or that of an individual tean member often encourages the others to perform better than they nor mally would.

It is important that a set of standards which embraces individua commitment, motivation and performance is set and communicate to the team members. The team members should be encouraged t meet these standards and to constantly strive to improve upon thei performance.

Support and recognition

External support and recognition are important to team success. Whe external support is missing team morale may suffer, especially if thos in a position of power are perceived to lack commitment to the tean and its goals. Senior management must therefore demonstrate its com mitment to team working by making changes in organizational struc ture and measurement systems. Providing structural support througl changes in performance review and promotion systems will encourag collaboration and help to generate higher levels of achievement. By realigning structures so that individuals become more dependent on the success of the team for their own success, the likelihood of intense loyalty to the team and the organization will be increased.

The most basic structural change involves tangible rewards. There must be a reward system that recognizes team effort and commitment to the team rather than purely individual effort. This involves developing a reasonable and relevant performance standard for the team and setting the compensation the team will receive for achieving the standard. The performance standard must reflect the level of performance that the team can be expected to reach if they focus on their objectives while working and behaving as a team.

Exercise 8.1

Generate a list of suitable standards that would facilitate the measurement of team performance in your organization.

Leadership

One of the most critical aspects of effective team performance is leadership. The team leader provides a focus on the team goals and is responsible for building an atmosphere of teamwork and cooperation among the team members.

Leadership involves the creation of a vision that is compelling and results-oriented, the communication of this to the team in a way that will encourage their commitment and the adoption of behaviour consistent with the vision. The leader must be capable of making things happen and must be able to motivate the team members toward achieving the objective.

Exercise 8.2

Think of someone who is recognized as an effective team leader in your organization. What qualities does this person possess?

Answer to Exercise 8.2

Your list of qualities is likely to include:

- The ability to avoid compromising the team's objective with political issues
- Personal commitment to the team's goals
- The ability to avoid diluting the team's efforts with too many priorities
- Fairness and impartiality to all team members

- A willingness to confront and resolve issues associated with poo performance from team members
- A willingness to be open to new ideas from team members.

The team leader must create a supportive decision-making climat that empowers people and creates enthusiasm. This can be accom plished by giving team members the confidence to take risks, mak decisions, and contribute to the team's success. It is also importan that the leader makes people feel part of the team and that the con tribution of the individual is encouraged. The team leader may also b responsible for the development of team members. Where this is th case, the team leader should develop a feedback mechanism that iden tifies performance levels and individual development needs.

Team development

As discussed earlier, the presence of the required technical skills anc personal characteristics in a team will be a major determinant of tean success. In studying the team roles that members can play and the skill most needed, Belbin's work is useful. However another method known as the Myers–Briggs Type Indicator (MBTI), is a powerful aic to team development.

The MBTI is based upon the work of Jung, who, in studying th underlying reasons for differences in human behaviour, suggeste that these were a result of personality preferences.[6] Jung's work were studied by Katherine Briggs and Isabel Briggs Myers, whc attempted to develop better ways to measure these differences. Thi work resulted in the development of an instrument known as th Myers–Briggs Type Indicator.

The MBTI reports preferences on four scales, each representing oppo sites:

Extroversion (E) or Introversion	(I)	–	where one prefers to focu one's attention
Sensing (S) or iNtuitive	(N)	–	how one acquires information
Thinking (T) or Feeling	(F)	–	how one makes decisions
Judging (J) or Perceiving	(P)	–	Orientation towards the oute world

It must be remembered that these eight labels reflect possible prefer ences. As an analogy, think of left- versus right-handedness. If a persor is right-handed, it does not mean that they cannot or do not use thei left hand. It simply means that they prefer the right. They may mak relatively little use of the left hand, or may be approaching ambidex terity. The same is true for the preferences listed above. A person may prefer one characteristic a great deal and another only slightly. Furthe examination of the two sides of each pair may indicate that the person identifies with both at different times, but within each pair there is one that will be preferred.

An individual's 'type' is the combination and interaction of the four preferences, and can be determined using a questionnaire. This type can be shown in shorthand by a four-letter code based upon the letter representing the preference. There are sixteen of these types in total as shown in Figure 8.1. As an example, ISTJ means an introvert who likes to process information with sensing, who prefers to use thinking to make decisions, and who mainly takes a judging attitude to the world. A person with opposite preferences on all four scales would be an ENFP. This is an extrovert who prefers intuition for perceiving, feeling for making decisions, and who takes a perceptive attitude to the world.

ISTJ	ISFJ	INFJ	INTJ
ISTP	ISFP	INFP	INTP
ESTP	ESFP	ENFP	ENTP
ESTJ	ESFJ	ENFJ	ENTJ

Figure 8.1 *The MBTI type table*

An analysis of an individual's type will reveal a dominant preference and an auxiliary preference, i.e. one function will take the lead and the other will help out. Both functions will balance and complement each other. However, it is important to realize that although everyone has a favourite function, all other functions are also available, and that for some purposes these may prove more useful. Indeed, third-favourite and least-preferred functions also exist for each of the sixteen types.

The following case, adapted from Kroeger & Thuesen,[6] provides an example of the use of the MBTI in an aerospace organization.

Case study

An ENTJ manager was experiencing difficulty in introducing a climate of collaboration rather than competition into his office. Power struggles were commonplace even at the expense of meeting office goals, and teamwork was practically non-existent. The manager decided to employ consultants to analyse the underlying reasons for these problems and to recommend solutions.

The MBTI was administered to all staff in the office to determine their individual differences. The group members' personality types were then plotted on the Type Table shown in Figure 8.2. The table

ISTJ 26%	ISFJ 3%	INFJ	INTJ 9%
ISTP	ISFP	INFP	INTP 14%
ESTP	ESFP	ENFP	ENTP 3%
ESTJ 23%	ESFJ 3%	ENFJ 3%	ENTJ 14%

Figure 8.2 *The group type table*

shows that there is an overabundance of some types and a shortage of others. The group was heavily TJ with Thinkers alone comprising 91% and Judgers 80%, of the group. Only three members of the group were Feelers, and, as is typical of any preference that is significantly underrepresented, they behaved unusually. One had a higher than average need for affirmation and a second was introverted. All three were considered ineffective and powerless.

A group discussion focused on issues of communication and interdependence, and particularly trust. Furthermore, the office manager was seen as someone who was a poor delegator, which did little to generate confidence among the office staff. By the end of the session two problems had been identified; the manager's ego and the lack of Feeling types whose position and sense of worth could be viewed by others as having leadership qualities.

The manager was a strong Thinker who was out of touch with his people-oriented Feeling side and therefore had difficulty in finding time for interpersonal issues on a daily basis. This led to high control needs which he used to compensate for things he did not do well, such as empowering his staff. In recognizing these preferences, and the presence of others who could deal with interpersonal issues with greater ease, the manager began to delegate more effectively.

The use of the MBTI in this particular organization showed that Thinkers and Feelers bring different needs and skills to the workplace, all of which are necessary for effective working together. Thinkers bring objectivity, clarity and productivity whereas Feelers bring an awareness of other people and their feelings and a need for harmony. All these qualities are essential to a successful team yet it is easy to block one half completely and to try to compensate for it by emphasizing the other half. Fortunately, such problems can be resolved relatively quickly using the MBTI.

In general, a knowledge of these preference types can be extremely powerful in understanding team behaviour when attempting to develop the team and when trying to improve team effectiveness. If, for example, a team member is forced to use only a favourite function without help from others, performance may prove to be adequate. However, when other less-preferred functions are used, performance may be superior. This could result in a team doing well in some areas while being unaware of their limitations in others – perhaps until it is too late. It is therefore important that the team is aware of its overall type and where it is over- or under-represented by a particular preference. By having a range of types in the team, each team member's contribution will be maximized and team effectiveness will be improved.

Conflict

Conflict can be defined as a process whereby an individual or group makes a determined effort to counter the efforts of another individual or group, resulting in the latter being frustrated in attaining its goals.[7] It can be both beneficial and harmful, depending on the type that exists. For example, if there is no conflict, complacency and stagnation may set in and performance may suffer as a result. Indeed, when controlled, conflict can increase motivation and innovation. However, too much conflict can lead to hostility and a lack of cooperation between team members. Therefore it is important that the types and causes of conflict are understood, and that conflict is properly managed.

There are two types of conflict: functional and dysfunctional. Functional conflict supports the goals of the team and improves performance, and is a constructive form of conflict. Dysfunctional conflict hinders team performance and is a destructive form of conflict. Robbins[7] suggests that an evolutionary process consisting of four stages leads to conflict outcomes, and these are described below.

Stage I: Potential opposition

The first step in the conflict process is the presence of conditions that create opportunities for conflict to arise. These sources of conflict are due to communication, structure and personal variables. Problems with the communication process such as insufficient exchange of information or overcommunication can give rise to conflict. Choosing the wrong channel for communication can also lay the foundation for conflict.

Team structure in terms of size and specialization can cause conflict. The larger the team and the more specialized its activities, the greater the likelihood of conflict. The adoption of a close style of leadership where continuous observation and restrictive control are in evidence or placing too much emphasis on participation and the promotion of differences can also cause problems.

In a team environment intra-team conflict can arise as team members become more dependent on one another or where strong personalities within the team disagree on something. Inter-team conflict can be caused by incompatibility of team goals or where there is competition for limited resources.

Stage II: Cognition and personalization

If the above conditions generate frustration then the potential for conflict will be realized during the second stage. The individuals or teams involved may become aware of the existence of these conditions and, as they become emotionally involved, this can lead to hostility towards one another. Unless preventative action is taken this will result in conflict.

Stage III: Behaviour

The third stage of the conflict process is reached when an intentional action is taken that frustrates the attainment of another's goals and the conflict is out in the open. Once the conflict is overt, the parties can develop a method for dealing with it. Thomas[8] has identified five conflict-handling approaches based on the degree of cooperativeness and assertiveness involved. These approaches and examples of when they should be used are listed in Table 8.1.

Conflict can be stimulated by placing teams or team members in competitive situations, and by bringing in outsiders to shake things up. It can also be stimulated by changing procedures, especially those that have outlived their usefulness. Conflict can be reduced or avoided by enhancing collaboration. Sometimes reaching a compromise is enough to resolve conflict. On other occasions, confronting the problem by bringing the parties together to discuss the nature of their conflict may be a more suitable approach.

Stage IV: Outcomes

The outcome of conflict will be either functional or dysfunctional. Conflict is constructive if it improves the quality of decision making through allowing all points of view to be heard. Where it stimulates creativity, provides a medium through which problems can be aired, or results in change it is also beneficial. Conflict is destructive where it breeds discontent, retards communication, or reduces cohesiveness. In extreme cases it can even threaten the survival of the team.

Table 8.1 The five conflict-handling approaches

Conflict-handling orientation	*Appropriate situations*
Competition	1 In emergencies where quick action is needed 2 On important issues where unpopular decisions have to be taken 3 Against people who take advantage of non-competitive behaviour
Collaboration	1 To find an integrative solution when both concerns cannot be compromised 2 To merge different perspectives 3 To gain commitment
Avoidance	1 With trivial issues 2 When there is little chance of satisfying your concerns 3 To let people cool down and regain perspective
Accommodation	1 When you are wrong and to show your reasonableness 2 To satisfy others and maintain cooperation 3 To allow individuals to develop by making mistakes
Compromise	1 To achieve temporary settlements 2 To arrive at expedient solutions under time pressure 3 As a back-up when collaboration or competition are unsuccessful

(Adapted from Thomas).[9]

The manager faced with excessive conflict should not assume there is one conflict-handling approach that will always be best. The technique chosen to resolve the situation should be appropriate to the specific situation.

Effective team management

In this and previous chapters a number of issues relevant to the successful implementation of a team-based approach have been discussed. Irrespective of the nature and role of the team, it is important that the team is managed in an effective manner. The remainder of this chapter focuses on an important aspect of successful team management: the team meeting.

All team meetings should be planned and organized in a manner that will ensure that they are conducted effectively and efficiently. For each meeting an agenda should be prepared and distributed to each team member in advance of the planned meeting date. The agenda should not be overloaded with an unrealistic amount of issues to be discussed. In addition, meetings should be held in a convenient location and at a time that will prevent other work activities from disrupting the meeting. A time limit should also be set for the meeting, with an agreed start time that is adhered to by all team members.

Meetings normally follow a set pattern that consists of six steps:

1 Review of previous minutes and the reason for and objective c holding the current meeting
2 Discussion, presentation and analysis of topics listed on the curren agenda including progress against plan
3 Evaluation and selection of solutions
4 Planning
5 Allocation of tasks, resources and timescale
6 Review and summary including setting a date, time and venue fo the next meeting.

Points discussed, decisions taken and actions arising from the meetin should be recorded in a set of minutes, together with the name of th team member tasked with the action and the agreed timescale. Minute should be:

- Concise and accurate
- In simple language
- Matched to topics listed on agenda
- Paragraph numbered for easy reference.

Although each team member should assume responsibility fo recording their specific actions, minutes should be circulated immedi ately after the meeting to ensure that there is a common understandin among the team of the actions required. The team leader can the monitor progress against the agreed plan.

Conclusions

The use of teams as a mechanism for improving organizational perfor mance is becoming widespread. However, few organizations are awar of the issues that have a direct bearing on the success of the team-base approach. In this chapter some of the fundamental issues relating to th use of teams have been introduced in an attempt to generate an aware ness of these and other issues relating to development of successfu teams.

There is a need to recognize the four phases of team developmen and the characteristics of mature teams. These issues have been dis cussed and the eight properties of effective teams have been identified The importance of clear goals which challenge the abilities of the tean and the selection of team members who possess the essential skills an abilities to accomplish the team's objectives have been discussed. Th need for collaboration and high standards, the common problems of lack of unified commitment, and external support and recognition hav also been identified. The need for a leader who is personally committe to the team's goal and who is prepared to give the team members the autonomy to achieve results has also been discussed.

Finally, a method that can be used when trying to develop teams and improve their performance has been presented. The different types of conflict that can be experienced in a team and the appropriate management responses have also been described.

Summary points

- The most common forms of quality improvement team are process teams and task teams.
- Teams pass through four stages during their development: undevelopment, experimentation, consolidation and maturity.
- The four important characteristics of teams are role structures, norms, cohesiveness and informal leadership.
- Successful teams have been shown to possess a range of common properties.
- Conflict within teams can be both constructive and destructive and must be properly managed.

References

1 Woodcock, M. (1989) *Team Development Manual*, 2nd edition, Aldershot: Gower.

2 Griffin, R. W. (1990), *Management*, 3rd edition, New York: Houghton Mifflin.

3 Larson, C. E. and LaFasto, F. M. J. (1989) *Teamwork: What Must Go Right/What Can Go Wrong*, Beverly Hills, CA: Sage.

4 Belbin, R. M. (1981), *Management Teams, Why They Succeed or Fail*, Oxford: Butterworth-Heinemann.

5 Anundsen, K. (1979), Building team-work and avoiding backlash: keys to developing managerial women, *Managerial Review*, February.

6 Kroeger, O. and Thuesen, J. M. (1992), *Type Talk at Work*, Delacorte Press.

7. Robbins, S. P. (1992), *Essentials of Organisational Behaviour*, 3rd edition, Englewood Cliffs, NJ: Prentice-Hall.

8 Thomas, K. W. (1976), 'Conflict and conflict management', in *Handbook of Industrial and Organisational Psychology*, Chichester: John Wiley.

9 Thomas, K. W. (1977), 'Toward multidimensional values in teaching: the example of conflict behaviours', *Academy of Management Review*, July.

9 Training for quality

Introduction

Education and training for managing quality is the essential foundation for success in establishing a reputation; a reputation as an organization which is determined to satisfy its customers. The way in which education and training are planned and implemented are as important as the content.

The purpose of education and training is to develop a sense of commitment to change supported by specific skills that enable the change to be initiated and sustained. In the context of quality management, the objectives must be related to the roles of every individual within the organization. Only one weak link in the chain has the potential to undermine the change process.

Education is the process of communicating a need – the need for a relationship with the customer which enables their requirements to be met. Training is important as the second element of interaction as it enables skills to be developed. In order to satisfy customer requirements, all individuals must be equipped with the necessary quality improvement tools.

The major difficulty in defining and carrying out education and training in quality management is defining what is meant by the term 'quality management' and by the other related subjects such as quality assurance, and quality control.

Consider again the basic definitions of quality.

Quality: The totality of features and characteristics of a product that bear on its ability to satisfy stated or implied needs.

Quality assurance: All those planned or systematic actions necessary to provide adequate confidence that a product or service will satisfy given needs.

Quality control: The operational techniques and the activities which sustain a quality of product or service that will satisfy given needs.

Clearly, these are general definitions and criteria which are meaningless unless given specific interpretations in specific contexts. The objective of education and training must therefore be to enable managers to interpret, and subsequently apply, training in the context of their circumstances in order to maintain and improve the quality of their activities. The objectives of this chapter are:

- To review the importance of a systematic approach to training
- To illustrate how the identification and analysis of training needs helps develop a training plan
- To review the different levels of organizational responsibility as regards training for quality
- To demonstrate how the quality-improvement tools represent an integral part of continuous improvement activity

The training process

Identification of training needs (ITN)

This is the first stage in formalizing an effective training programme. The identification of training needs should be carried out in two main areas:

- Problem analysis
- Performance analysis of departments and individuals.

Problem analysis

This should be done by identifying problems being experienced by the company. Examples of problems are:

- High reject rate
- High level of customer returns
- Missed delivery dates
- Delays with introduction of new products
- Drop in sales
- Poor service to customers.

The next step is the analysis of the problems to indicate the major causes. For example, is the high reject rate and high level of customer returns due to manufacturing faults, to design faults or to defective raw materials? If it is a manufacturing fault, is it primarily related to the operator or the equipment? Are missed delivery dates or the introduction of new products due to bad productivity, poor production planning, or products being reworked or replaced due to poor quality?

Is drop in sales due to missed delivery dates, poor quality, high price, increased competition, a declining market, or an out-of-date product range? From another viewpoint, accepting that organizational targets for productivity and quality have been achieved, may cause one to overlook that the performance of one department at 105% of target has compensated for two other departments' performances at 98% and 97%. Similarly, accepting that one department has achieved its target may conceal the fact that some operators are compensating for othe operators with regard to either productivity or quality performance.

Generally, operator faults fall into three categories:

1 Inadvertent faults which are quite infrequent and are caused by lac of concentration or carelessness, and can only be eliminated com pletely by foolproofing the operation. However, they can b reduced by identifying responsible operators and by better supe vision and training
2 Technique faults are normally recurring faults, either with one c more operators, and can be reduced by the training of operator However, what are frequently identified as operator-related faul are misrepresented. This is particularly so in the area of qualit failure, where a frequent difficulty occurs with different interpreta tions of the standard of quality required. To eliminate these prob lems, all employees, not just operators, must have a commo understanding of quality and understand their personal role i causing quality to be routine
3 Conscious faults can be witting, intentional and persistent. A change in behaviour is required to overcome these faults and trair ing may have a role.

Once the cause or causes of problems have been identified, the ne step is to identify needs. These could be of a training or non-trainin nature. For example, a high reject rate identified the following needs

1 The introduction of raw material inspection (non-training need)
2 The training of inspectors in the new raw material inspection prc cedures (training need)
3 The modification of quality standards for operators (non-trainin need)
4 The training of operators and inspectors in these new standard (training need).

The example of missed delivery dates may identify the followin needs:

1 The introduction of new equipment (non-training need)
2 The setting of production standards for each operator (non-trainin need)
3 The training of operators in the new methods of productio (training need).

Performance analysis of departments and individuals

To ensure an adequate identification of training needs, it should b approached on a top-down and a bottom-up basis. The reason is tha minor variances on a department-by-department basis may not b significant for the individual department, but the cumulative effect for the company could be very significant. Examples would be a department having an unfavourable yield variance of 0.5, which may not be noteworthy for that department but similar results for six departments give a variance of 3%. This could be critical from the company's viewpoint. Similarly, a production delay of four hours in each of six departments may result in missing a delivery date by three days.

Checklist for identifying training needs

1. At the outset of the ITN, clear terms of reference should be determined and agreed by management. Put them in writing and give a copy to management. These are your agreed and written objectives. If it is found necessary to change your terms of reference, inform management, rewrite them, and give management a copy of the new terms.
2. Inform people about what you are trying to do. Nothing is more suspicious in the work environment than asking people searching questions about their work without explaining why you want to know. All information obtained should be regarded as confidential.
3. Determine the range of work and the degree of versatility required – e.g. are there any special learning difficulties encountered by the operator?
4. Identify the common skills.
5. Determine the areas where job knowledge is involved. Is there sufficient job knowledge for the operator to work efficiently?
6. Establish the present operator performance. Closely examine the reasons for operators not making Experienced Workers Standard in a certain time – or at all.
7. What is the organizational set-up? How is the work flow controlled? Any hold-ups due to shortage of materials – is quality control, inspection or test involved?
8. Obtain a quality specification if there is one. Determine the overall quality standards. What is the scrap/rework rate? How much does it cost?
9. What are the production control objectives and output figures?
10. What happens to the documentation? What is its purpose? Who uses it and why? Is it being used effectively?
11. Are the setting-up methods and machine conditions adequate for the tolerances required? Is the preventative maintenance adequate? How much downtime is there? Can any corrective action be taken if inefficient running is discovered?
12. Has the working environment an effect on the quality or quantity of production output? Remember that the training environment should match the actual working environment as closely as possible.

13 Accident rates should be investigated and the reasons for accident sought. Reportable accidents should be looked into very closely How do the figures compare with local and national averages?
14 Does the company have any known plans to expand generally, or i a specific department, or, even set up a new kind of department There may be a training need if the answer to any of these question is 'yes'.
15 It is very important to estimate, as accurately as possible, cos savings which result directly from the application of systemati analytical training. For example, we should investigate the cost o present training arrangements, the cost of proposed training an the saving in rework or scrap or in any other area due to bette training and state the saving in pounds.

Analysing training needs

Information required

The purpose of analysing training needs is to determine what type o training is required for specific operations and what type can make th most economic contribution towards the solution of a problem. Certai information needs to be gathered so that an analysis can be made of th most appropriate type of training required:

1 *Range of work*: In many tasks this is likely to present a problem o variety in the sense that the experienced worker is capable of pro ducing a wide range of products; operating several types o machine or using a range of materials. These abilities have ofte been built up over the years of experience and it is necessary t discuss with management whether learners should be trained t achieve this level during their initial training course. It must b remembered that there is no point in acquiring a skill if the traine is not given an opportunity to use it within a reasonable perio under production conditions. If this is not possible, the skill i likely to be lost and may have to be relearnt. On the other hand the problem of variety of product often appears more complex tha it is in the sense that skilled operatives have learnt to use a limite range of basic skills and can apply them to produce a wide range o products.
2 *Length of present training and cost of training*: When considerin length of present training it is important to ensure that the infor mation given relates to the time taken to reach average experi enced worker's standard of quality and output. Training times t earn basic wages take no account of the cost of loss of overhea

recovery. Present training times can be ascertained by examining wages or production sheets.

3 *Present training arrangements*: Some typical questions to ask are:
   - How long do new starters take to reach experienced workers' standard?
   - Who is responsible for training new employees?
   - How is the training carried out?
   - Have instructors been trained how to instruct?
   - What checks are there on trainee progress to evaluate suitability, output and quality?

   It is also necessary to assess the attitude of supervision to training, and to discover whether they recognize a need to improve the training arrangements.
4 *Organization of department*: The general background to production in the department must be ascertained to find out such matters as:
   - How flow of work is organized
   - How work is inspected and quality assured
   - How operatives receive instructions
   - How machinery is serviced.
5 *Analysis of future company plans*: Some estimate of future production requirements must be made in order to forecast the number of trainees required. Because analytical training reduces training times, the number of trainees required is reduced as fewer people are in 'training' at any one time. Thus a planned expansion of production might be achieved with no increase in numbers in the labour force. Other factors to be taken into account include:
   - What technological changes are envisaged, i.e. new skills required?
   - If a new factory is being opened, will the local labour force possess the necessary skills?
   - Will any changes in product demand new skills of operators?

The training plan

The information obtained during the training needs analysis acts as a basis for the training plan. This should cover such matters as:

1 A precise statement of the objectives for the training detailed in quantifiable terms where possible
2 The nature and methods of the training to be adopted. For example, there may be formal in-company courses or planned experience on the job
3 The duration of the proposed training periods
4 The priority of each proposed training item – for example, immediate, short or long term, as part of an overall training plan

5 The cost of the training where it is practical to estimate this
6 Target dates for completion of the specified training activity.

Ongoing nature of training

The investigation of training needs should become an ongoing activit and not merely a 'once-off' exercise never to be repeated. The mai reasons for this are that the marketplace is continuously changing. A companies are subject to change (as a result of market changes c internal pressures) and consequently their training programmes nee to be updated.

As training expertise in the company develops, more in-depth need should be identified. Figure 9.1 indicates the total process. Followin

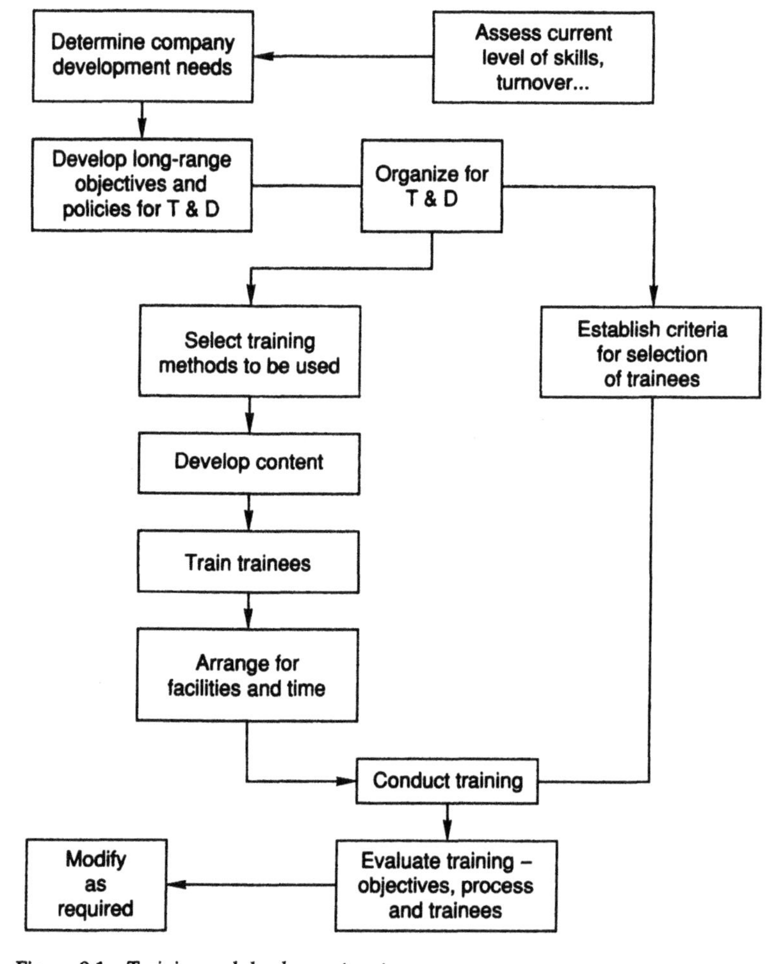

Figure 9.1 *Training and development system*

an initial identification of training needs, a training plan is produced. The training is carried out and the effectiveness of the training is evaluated. Results of the evaluation, in addition to changes which take place within the company, will necessitate that the cycle starts again.

Evaluation of training

Evaluation is the assessment of the total value of a training system, training course or programme in social as well as financial terms. Evaluation of training is performed at a number of levels and may have to use a range of methods. Hamblin[1] gives five levels of effects of training which can be evaluated:

1. *Reactions*: This concerns the subjective valuation of the training course by the trainees in terms of their impressions, opinions and attitudes. Such reactions may be a function of the training content and methods used; its location; other trainees; the trainer, etc. The reactions will be there, regardless of the trainer's intentions, and on some occasions may affect the attainment of the behavioural goals of the training programme. The trainer must therefore attempt to identify the reactions which may inhibit not only the internal but also the external validation of the programme. Interview and questionnaire techniques are useful methods for collecting information about such reactions.
2. *Learning*: Training research has been mainly concerned with the evaluation of learning, following different training methods. The learning/performance distinction is important here, as this emphasizes that learning can only be indirectly inferred from performance measurement. The main criteria for the selection of appropriate measures of performance must be their validity and reliability. That is, they actually measure relevant dimensions of training and that their results are repeatable. Tests of training at this evaluation level have similar requirements to psychometric tests. Evaluation here is of the efficiency of the training programme rather than of the trainees and any failure must be attributed to 'faults' in the training process.
3. *Retention*: Retention of training is particularly important where recently acquired skills are not practised and may be forgotten. Some factors affecting retention are the nature of the task and how it is measured; the amount of training; the duration of the retention interval; and the nature and time of occurrence of any other activities performed before or after training.
4. *Job behaviour*: The training programme must not only be internally valid by achieving its behavioural objectives but also externally

valid with adequate job performance. The level of training manifest ing the job situation is determined by the amount of 'transfer of training'. Measurement of transfer of training can be fairly straightforward for human–machine tasks, but may be more elusive in other jobs (e.g. those requiring social skills such as customer service or supervisory training).

5 *Organization*: Hamblin suggests that training may also be evaluated in terms of changes in the functioning of the organization for which Lindahl[2] has presented eight criteria:
   (a) Quality of production
   (b) Number of operators able to reach job standards
   (c) Time required for a specific job
   (d) Damage to material or equipment
   (e) Absenteeism
   (f) Labour turnover
   (g) Running costs
   (h) Performance on personnel measures such as tests, rating scales and attitude surveys.

Evaluation of training is a complex issue which may proceed at many levels. However, the system viewpoint illustrates how evaluation is an essential feature of the organization of training. Efficient control of the system is only possible by the *feedback* which evaluation provides. The system can therefore not only correct its own errors but can also be adaptive to changing circumstances. Thus, the same underlying principle of feedback control which is useful for describing human performance also provides a useful framework for describing the regulation of training systems (see Figure 9.2).

Organization responsibilities for training in quality

What is the context of education and training for managing quality? It is one part of the enabling process for success in an organization's effort to be dedicated to customer satisfaction. Thus it is necessary to focus on the overall process of improvement and ensure that trainees at all levels relate to an environment in which investment of time and effort in learning is given long-term value.

What training and education needs do the various levels of employees have in order to make them effective enablers of improvement? Typically, five levels of responsibility can be identified for most organizations:

1 Senior management
2 Middle management
3 Line supervision
4 Team leaders
5 The individual

Compared to two years ago	(Check one) Yes	No	?
1 Does your manager have a better understanding of how you perform your job?			
2 Does he have a better understanding of you as an individual?			
3 Does he better indicate recognition of your good work?			
4 Does he better utilize your particular skills?			
5 Do you have a better picture of what he expects from you in terms of job performance?			
6 Do you have a better picture of how you stand with him overall?			
7 Does he discuss your job performance with you more frequently?			
8 Do you have a greater opportunity to present your side of a story during those discussions?			
9 Does he take a greater personal interest in you and your future?			
10 Does he make a greater effort to help you develop yourself?			

Figure 9.2 *Example of a behavioural questionnaire (adapted from Moon and Hanton[3])*

Senior management

Recent evidence indicates that organizations succeed or fail in their drive for positive management of quality in direct proportion to the amount of visible commitment from senior management level. Their responsibilities include:

- Harnessing quality-improvement activities to the corporate goals of the organization
- Ensuring that learning needs at individual, departmental and overall organizational levels are identified
- Ensuring that a purpose, policy and plan for quality development are established
- Ensuring that adequate resources are available for the operation and evaluation of the plan.

Thus, the education process for the senior executive needs to stress the imperative nature of personal commitment, continually reinforcing the messages and ideas that need to be heard at all levels (see Figure 9.3).

Senior management training

- Introduction to total quality. Emphasis on the competitive environment and the need to use quality as a differentiator
- The need to own a strategy and vision for quality

	Executive senior/ management	Middle management	First-line management supervisory level	Quality improvement team leaders	Other employees
Quality improvement as a corporate strategy	→				
Leadership skills and strategy formulation	→	→			
Organization and planning the improvement process	→	→	→		
Individual role and tools and techniques for improvement	→	→	→	→	

Figure 9.3 *Matrix of education and training planning requirements*

- Development of an action plan
- Appreciation of the tools and techniques for quality improvement.

Middle management

Typically, middle managers carry the functional responsibility for ensuring that employees are helped to perform their jobs effectively and efficiently. Their responsibilities include:

- Guiding the transformation process as part of a steering committee and/or quality improvement team
- Planning implementation
- Leading work groups in the use of measurement and problem-resolution techniques

The skills to be developed are primarily concerned with putting the quality improvement process into practice and carrying out quality improvement team tasks.

Middle management training

- Introduction to total quality
- Development of an implementation plan for local areas
- In-depth study of tools and techniques of quality improvement.

'Middle manager' covers a wide spectrum of responsibility. In larger organizations, the training programme for middle managers who are at the higher end of the spectrum will have much in common with the programme for senior management. At the lower end of the spectrum the training may well overlap with that used for first-line supervisors. In planning training activity, consideration should be given as to whether or not it is preferable to offer a single standardized programme for middle managers or to design specialized programmes for the various categories of middle managers.

First-line supervisors

This category of management has traditionally been exposed to training courses which have emphasized:

- Elements of supervision
- Work planning
- Job technology (processes, materials, tests, etc.).

Recently the quality-oriented content has become more prevalent primarily due to a shift in emphasis as regards the role of first-line supervisors/junior managers in quality-improvement activities which emphasize the team effort in solving problems.

First-line training

• What is quality control?	definitions; history; how to control quality
• Role of the supervisor:	what to control; guidelines; procedures
• Pareto Diagram:	how to construct, interpret and use
• Ishikawa Cause and Effect Diagram:	how to construct and use
• The statistical viewpoint:	importance of the factual approach; setting up to collect data (data sheets, recording); frequency distributions
• Control charts:	how to make and use (both variables)
• Process improvement:	how to analyse processes; improve them and establish controls
• Sampling concepts:	random sampling; sampling tables; sampling error
• Application:	how to apply the tools in the shop environment
• Quality circle activities:	organization, training and working with quality circles

- Quality assurance: establishing procedures; developing monitoring mechanisms; performing audits; ensuring corrective action is carried out.

For further detail on control charts, sampling concepts, etc. se Chapter 5.

Team leaders

As team leaders may well hold middle, junior or first-line managemen status, there is potential for overlap in the training provision. While

- An in-depth working knowledge of the total quality programm and,
- A working knowledge of quality improvement tools are necessary emphasis should also be placed on specific skills for leading tean activity:
  - Communication
  - Presentation
  - Interpersonal
  - Teambuilding.

The individual

It is at this point in moving down through the organizational hierarch that consideration has to be given to how a company 'manages' qual ity. As the efforts of all employees are to be geared towards managin individual responsibility for quality, it is important to determine wha it is that management expects of its employees.

Individual quality accountability rests equally with everyone. Th most effective demonstration of commitment of a senior manager is preparedness to show personal determination by applying quality improvement tools and techniques to their own job. All employee need to be able to identify their customers and suppliers and to asses the extent to which their requirements are understood and satisfied.

It is a significant challenge to communicate consistent messages i training and education for quality management to all employees. I requires:

- An objective view of the individual value set to be implemented
- Tools to be used so that communication between individuals i unhindered
- An introduction to common systems to support personal and tean efforts to effect quality improvement.

This process could be regarded as one of empowerment; removing th traditional barriers to improvement where individuals feel that an suggestion for change would be best in an environment where pressures of cost and schedule override quality. Training for all employees should be such that they are fully aware of:

- Why the quality programme exists
- Who is responsible for quality
- The objectives of the programme
- Individual roles in an organizational context.

ools and techniques of quality improvement

Once management have been convinced of the need for change and all employees made aware of the reasoning for the change initiative it is essential for everyone to have the relevant knowledge or ability to use basic problem-solving tools.

Flow charting

A flow chart is a diagram that shows all the major steps of a process. Preparing a flow chart is one of the first things to do in analysing a process. It uses a set of standard symbols to document the process steps, presenting them in a pictorial format that is easy to understand.

Through flow charting, team members can better understand the processes for which they are responsible. The flow chart demonstrates how the different steps in a process are related to each other. It provides insight for identifying value-added activities, control points, data-collection points, inefficiencies in the work flow and obvious key points in the process. It is also an excellent training tool for new employees. A flow chart is used:

- To analyse relationships between sequential activities
- As a technique for fully understanding a problem
- As a source of information for problem identification and resolution
- To analyse customer or supplier activities.

How to flow chart a process

1. Bring together representatives from all departments responsible for the process so they can perform the analysis together
2. Title the chart with the name of the process analysed. If there is more than one, diagram them on separate charts and number them sequentially
3. List sequentially all major steps involved in the process. In some cases it may be easier to start at the end of the process and work towards the beginning. However, the flow is always shown

beginning at the top-left corner of the chart. Make sure that proces
boundaries are clearly defined

4 Using the set of symbols shown in Figure 9.4, draw a flow diagran Concentrate on major processes so that the flow chart will fit on single page if possible. The chart should represent the way thing are, not the way they are supposed to be

5 When processes are complex, create second- and third-tier flov charts as necessary to adequately break down all major processe into the component parts (see Figure 9.5).

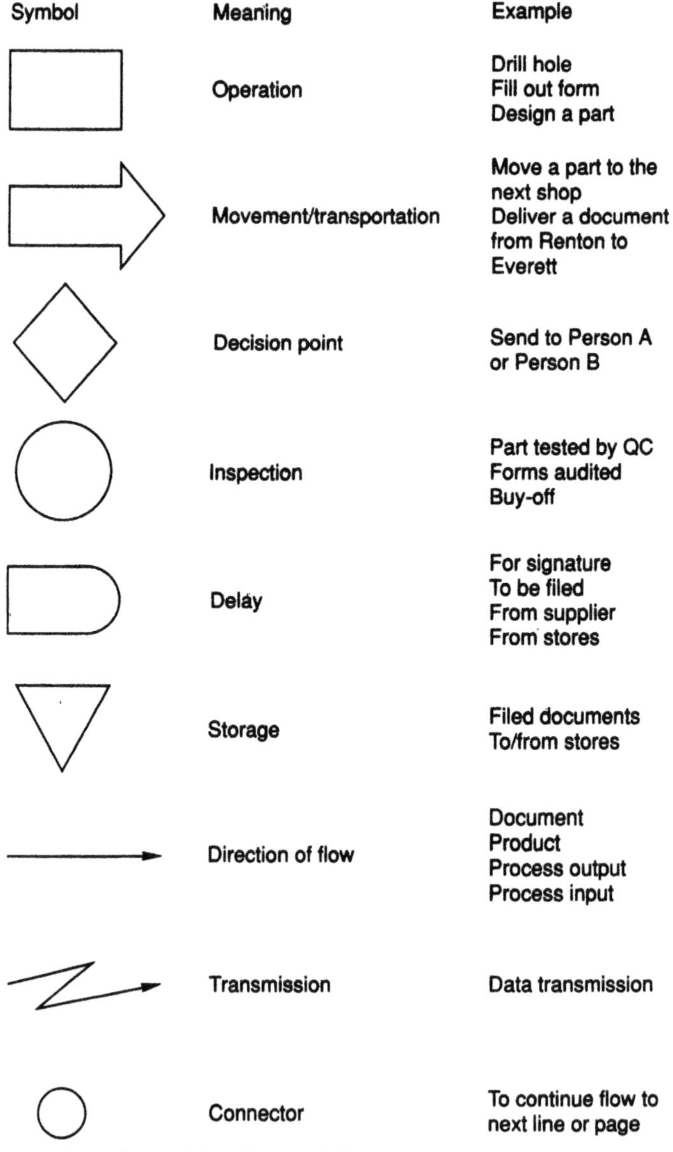

Symbol
Meaning
Example
Operation
Drill hole
Fill out form
Design a part
Movement/transportation
Move a part to the next shop
Deliver a document from Renton to Everett
Decision point
Send to Person A or Person B
Inspection
Part tested by QC
Forms audited
Buy-off
Delay
For signature
To be filed
From supplier
From stores
Storage
Filed documents
To/from stores
Direction of flow
Document
Product
Process output
Process input
Transmission
Data transmission
Connector
To continue flow to next line or page

Figure 9.4 *Standard flow chart symbols*

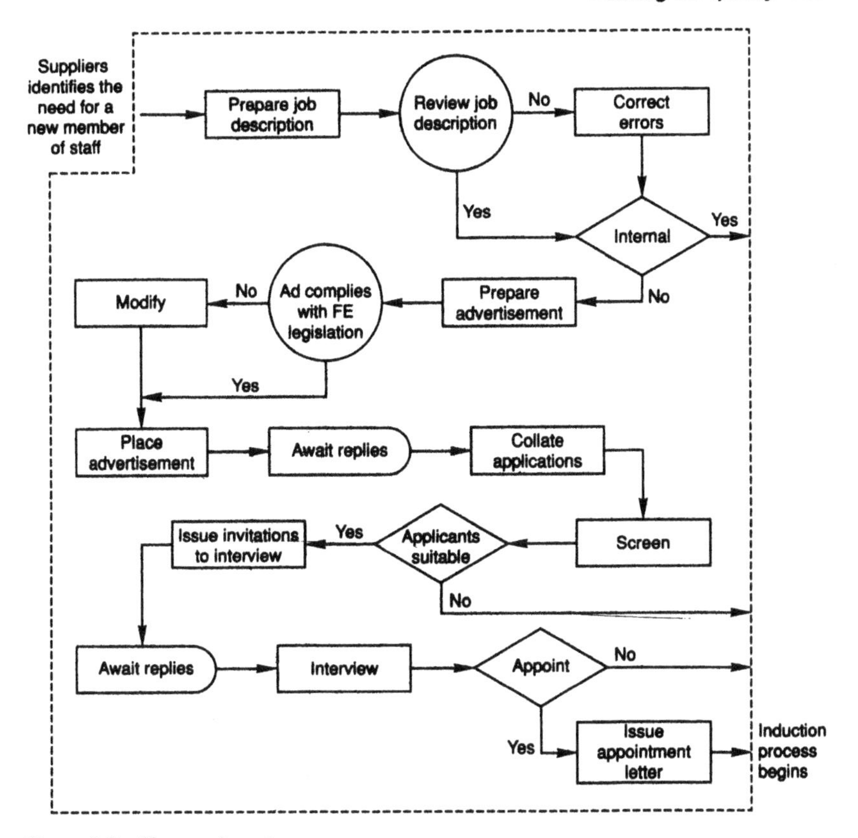

Figure 9.5 *The recruitment process*

Brainstorming

Brainstorming is, without doubt, the most fundamental of all problem-solving or analytic tools. It is a way of generating a large number of ideas from a group of people in a short period of time. An individual would undoubtedly find it difficult to create a large quantity of ideas on a particular topic but thinking collectively, a group can produce many thoughts, each idea leading to others, initiating tangential thoughts and, in many cases, finding solutions to problems.

Despite what it may seem, brainstorming requires a somewhat structured approach:

- The process should take place in small groups – usually between four and eight persons. Having the group too large or too small can, in fact, inhibit participation
- The membership of the group is a very important area and should be dictated by the topic to be 'brainstormed', i.e. all members from the same work area for localized problems or perhaps a cross-functional team for more strategic issues

- Each group requires a leader to keep the process focused on the topi area, to promote thought and to record the outputs
- Finally, the process should take place in a no-blame, no-interruption environment.

There are some basic brainstorming rules which, if applied, will assis the leader and team members to achieve the maximum number of idea in the limited timeframe available.

Rules of brainstorming

- Define and write out the topic
- Take turns of offering ideas in sequence
- Maximize the quantity of ideas generated
- Wild ideas are welcome
- Do not criticize people's ideas
- Do not discuss ideas
- Build on ideas
- It is 'OK' to pass
- Record information as given.

Having generated as many ideas as possible within the timeframe, th next step is to discuss the relative merits of each idea. Some ideas o the list will merely be gripes, others may have great value. It is impor tant to be able to prioritize these ideas in some way, as it would not b feasible to address each idea and continue with the day-to-day runnin of the business. Many prioritizing techniques exist, including prioritiz ing by criteria and prioritizing by paired comparisons, but perhaps th most well known is Pareto Analysis. This is based on the Pareto effect more often referred to as the 80-20 rule.

Pareto Analysis

Pareto Analysis is a tool used to prioritize problems for solution whicl highlights the 'vital few' from the 'trivial many'. The vital few are th factors accounting for the largest percentage (80%) of the total, whil the trivial many are the myriad factors that account for the smal percentage (20%) remainder. The 80-20 rule suggests that approxi mately 80% of the value or costs comes from 20% of the elements, e.g

80% of the sales comes from 20% of the customers, or
80% of the cost of inventory is tied up on 20% of the parts.

Pareto diagrams are used mostly in conjunction with collected numer ical data or with ideas produced by brainstorming. They are essentiall bar charts with a unique feature which is that the bars are placed i order of importance – usually cost or frequency.

The following example will highlight use and application of Pareto Analysis. Consider the data sheet of common error in any administration department (see Table 9.1). The first two steps towards constructing a Pareto chart are:

1 Determining the items to be investigated. This can be done by brainstorming
2 Designing an appropriate data collection sheet. (This has already been done in Table 9.1.)

Having collected the data, a frequency table needs to be produced. This highlights the category of error, the frequency of its occurrence or cost, the cumulative frequency, the relative frequency, and the cumulative relative frequency. Thus, Table 9.1 may be transformed into Table 9.2. In this case, the 'other' category occurs 1.9% of the time. If the others account for 50% or more of the data, then the breakdown of categories must be reformulated. Table 9.2 shows that three of the categories account for approximately 80% of the total errors. To show this more clearly, we construct a Pareto diagram.

Table 9.1 Record of defects for keypunch operator

Major causes of defective cards	*Month*				
	1/88	*2/88*	*3/88*	*4/88*	*Total*
Transposed numbers	7	10	6	5	28
Off-punched cards	1		2		3
Wrong character	6	8	5	9	28
Data printed too lightly on card		1	1		2
Warped card	1	1		2	4
Torn card			1	1	2
Illegible source document			1		1
Total	15	20	16	17	68

Table 9.2 Frequency table of defects for keypunch operator

Major causes of defective cards	*Frequency*	*Relative percentage*	*Cumulative frequency*	*Cumulative percentage*
Transposed numbers	28	41.2	28	41.2
Wrong character	28	41.2	56	82.4
Warped card	4	5.9	60	88.3
Off-punched card	3	4.4	63	92.7
Data printed too lightly on card	2	2.9	65	95.6
Torn card	2	2.9	67	98.5
Illegible source document	1	1.5	68	100.0
Total	68	100.0		

Construction of a Pareto diagram

1 Draw horizontal and vertical axes on paper and add the appropri ate unit, i.e. for the vertical scale, the frequency or cost
2 Under the horizontal axis add the most frequently occurring o most costly category on the extreme left and continue in decreasin order to the right
3 Draw in bars representing the value of each category. Sometime this bar chart alone may be enough to base decisions on. Howeve there will be cases when the cumulative percentage of adjacent bar will be needed
4 Plot the cumulative line (cum line) on the Pareto chart. This is don by starting at zero on the diagram and moving diagonally to th top-right corner of the first bar. This process is repeated adding th number of observations in the second bar and so on until the lin reaches the total number of observations.

Pareto diagrams must describe when and under what conditions th data were gathered if they are to be useful for comparisons and ascer taining whether change has taken place.

Figure 9.6 shows a plot of frequency; sometimes it is more forceful i cost is placed on the vertical scale. In this case, the cost per error o

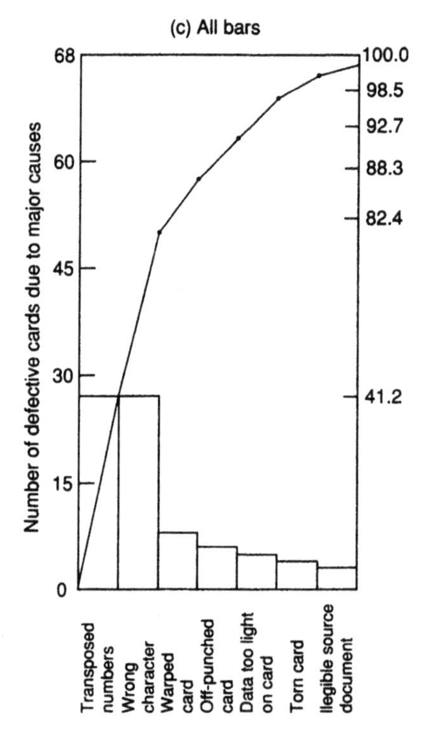

Figure 9.6 *Cumulative percent of major causes of defective cards*

occurrence must be calculated. This might require the order of the errors on the horizontal scale to be altered if, for example, the total cost associated with incomplete documents were greater than that of typing errors.

Having identified the major problem areas using Pareto Analysis, corrective action must be taken to address these areas. Then, collecting more data on the same department should show a change in Pareto charts. Again the vital few may be identified and corrected. In this way, continuous improvement takes place.

Cause-and-effect analysis

The cause-and-effect diagram is attributed to Dr Kaoru Ishikawa[4] (Figure 9.7). From his understanding of the problems faced every day by plant engineers, he realized that the most complex task they faced was in coping with the multitude of factors affecting their processes and solving their problems. Cause-and-effect diagrams, however, are not limited to use by plant engineers. They have widespread application by all members of staff whether in manufacturing or service industry.

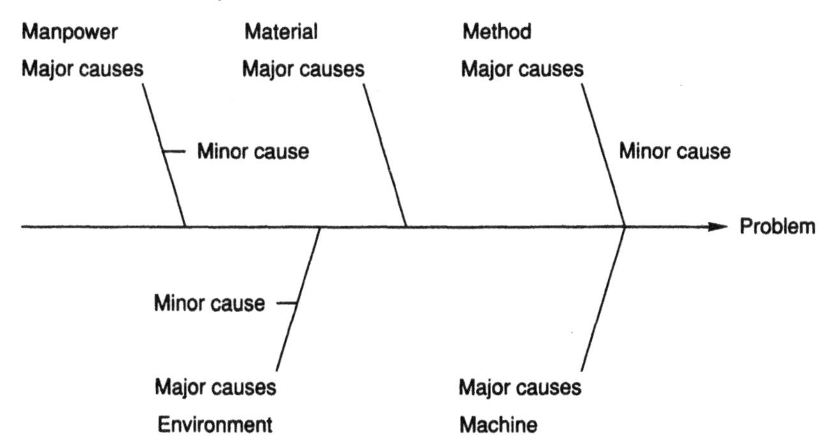

Figure 9.7 *Cause-and-effect diagram*

Cause-and-effect analysis tends to be used, after brainstorming, to organize the generated information. It is useful in helping us to understand the root causes behind our problems. The technique consists of defining a problem (or effect) which occurs in the workplace and needs to be changed or corrected. Once this effect is defined, the factors contributing to it are required (causes). While there are possibly only one or two causes of a problem, there are probably many potential causes. These should appear on the cause-and-effect diagram for later discussion.

When we think of any process and the factors which affect it, we should think of the following:

- Man
- Machine
- Method
- Material
- Environment.

These are the inputs to a process and, as such, obviously affect it output. Thus, errors at the inputs could be the causes of the prob lems, i.e. the effects.

The generic fishbone diagram

There are generally six steps to preparing a fishbone diagram:

1 Gather a work team together as for brainstorming
2 Identify the problems to be examined
3 Draw the generic fishbone diagram as a flipchart and include th identified problems as the *effect*
4 Brainstorm for probable causes under each heading:
   - Man
   - Machine
   - Method
   - Material
   - Environment
5 Ascertain the most likely cause, either by using past data or by group consensus
6 Verify the cause by data collection.

Points worth noting

- The five main generic headings may not be sufficient or relevant s change them accordingly
- Too many major headings imply lack of knowledge about th process
- Listing the causes is only the first step to solving the problem
- Fishbone diagrams are useful if you are looking at an individual par of an overall process.

Process Analysis

Process Analysis is another form of cause-and-effect diagram. It is use when a series of events, perhaps multiple steps in a process, gives ris to a problem and it is not clear which step gives rise to the problem.

- In this case, as before, the problem must be defined. The problen could be a defect in the output of an assembly line, for example

- Each stage of the process must be examined for possible causes of the problem. This can be done by brainstorming at each stage
- When each significant cause has been selected, it must be verified by data collection or experiment
- Process analysis is ideal for structured processes which can be laid down in a step-like fashion.

Scatter diagram

The fishbone diagram provided a simple means of relating causes and effects. The scatter diagram is a graphical method of determining the relationship between the cause and effect through pattern analysis.

Scatter diagrams are used to assess whether there is an association or correlation between two process parameters (for example, the hardness of a painted surface in an oven and the temperature of that oven). If one can control the temperature accurately, one can also control the hardness of the painted surface.

Plotting the hardness of the surface against the oven temperature will probably not result in a straight line but rather a series of points scattered about some central line or curve. It is the pattern in which these points lie which determines the relationship, if any.

Steps to constructing a scatter diagram

1. Identify dependent and independent parameters and take measurements – 50-100 would be useful
2. Draw the axes and plot the points. The independent factor should be on the horizontal axis
3. Draw a best-fit line through the points
4. Analyse the resulting scatter diagram.

Histograms

Histograms may be thought of as bar charts which show patterns of variation. They are, like Pareto charts, a graphic representation of frequency tables. Histograms are created by dividing raw collected data into equal intervals. The number of measurements falling into each interval is counted and bars are then constructed so that their heights are proportional to their frequency of occurrence.

The histogram thus produced graphically illustrates three characteristics of these raw data:

- The first is the central tendency or nominal. This value is usually called the average. It is the value around which the data are predominantly clustered

- The second characteristic is the range. It is a measure of the tota[l] spread of the distribution of data and is usually measured in terms o[f] standard deviation
- The third characteristic is the shape of the data. Usually data fall int[o] a bell-shaped curve called the 'normal distribution'.

Steps to constructing a histogram

1. Identify and define data to be collected
2. Construct a data-collection sheet
3. Collect data
4. Locate smallest and largest measurements
5. Calculate range
6. Select a number of intervals
7. Determine class interval size
8. Determine class limit end points
9. Tally measurements by class intervals
10. Draw bars and labels.

Drawing a smooth curve through the bar chart also helps in determin[-]ing whether the curve is bell shaped.

Histograms may be used diagnostically as follows:

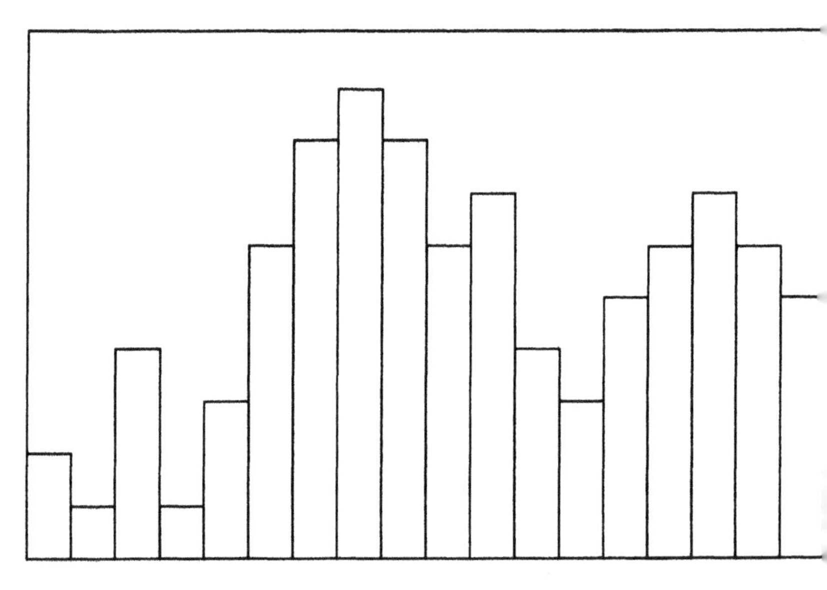

Figure 9.8

- Figure 9.8: bimodal distribution. This shows that multiple processes are at work and usually means different data sources.

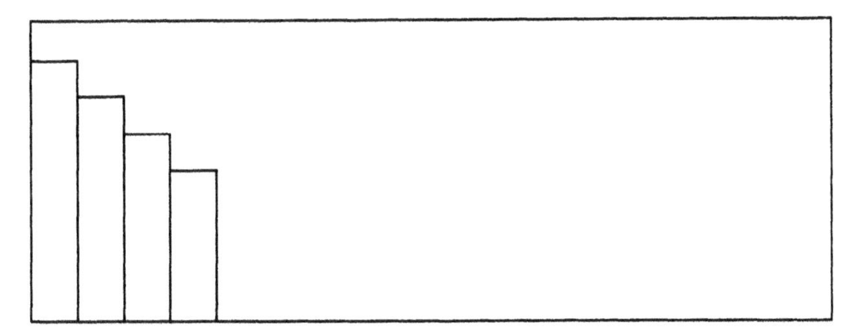

Figure 9.9

- Figure 9.9. Peak not centred. This could mean high scrap rate in production.

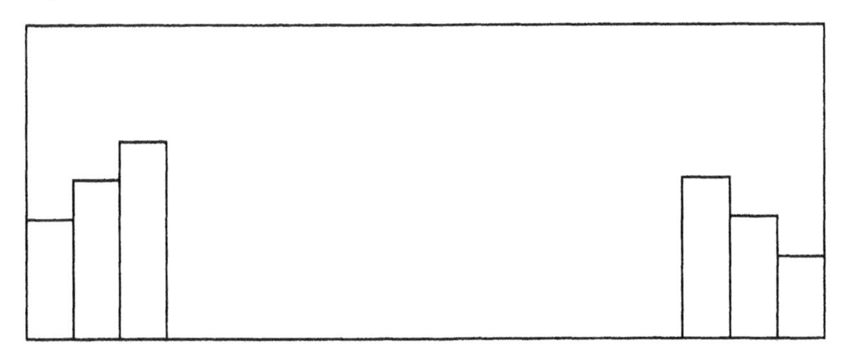

Figure 9.10

- Figure 9.10. This distribution shows the data or items around the centre have been removed. It could indicate that the tight-tolerance products are being kept for another customer.

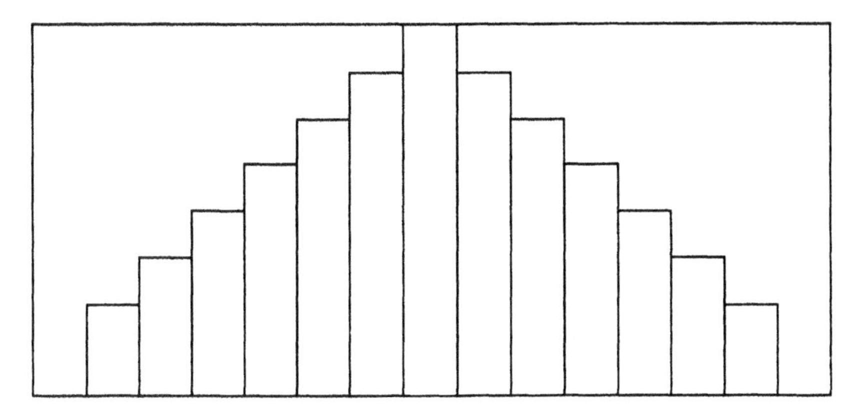

Figure 9.11

- Figure 9.11. Normal distribution.

Exercise 9.1

The following data describe the internal diameter of shims ranked in order of size (mm). Plot a histogram of the data and, given the specification of 23 ±1.0 mm, comment on the distribution.

Shim size (mm)				
23.3	23.5	23.6	23.7	23.8
23.3	23.5	23.6	23.7	23.8
23.4	23.5	23.6	23.7	23.8
23.4	23.5	23.6	23.7	23.8
23.4	23.5	23.6	23.7	23.8
23.4	23.5	23.6	23.7	23.9
23.4	23.5	23.7	23.8	23.9
23.4	23.6	23.7	23.8	23.9

Case study: quality improvement project

Project: Reduction of customer back orders

Leader: Operations manager – customer support operations (CSO)

The team was chosen due to individualistic strengths within various disciplines which are influential in supporting spares. The interaction between areas was also deemed essential, i.e. the engineering and customer support group and the operations group.

Team Members: Material planner
Progress controllers – CSO
Warranty administrator – CSO
Team leader/production engineering

Setting the scene

1 *Measurable objectives*: At the commencement of the project, August 1991, £0.5million was outstanding in terms of firm customer orders. Furthermore, £170 000 was outstanding six months plus. The objective of this project was to make a significant impact on monies outstanding. The benefits of the project may be regarded as two-fold:

- Tangible – monetary return
- Intangible – improved customer care and benefits derived from same.

The supply of spares emanates from essentially two sources:

- Outside vendor
- In-house – made-on-works (MOW)

Although the project encompassed both areas, the major concern was in-house supply, with specific attention to aircraft division operations.

Project commencement 145K 6 months+
Project completion 90K now measured 3 months

2 *The manufacture of spares – culture*: In order to fully comprehend the task ahead, a brief insight into the culture/attitude towards the manufacturing of spares is essential. The manufacture of spares has historically taken low priority within both production and pre-production support activities. This is not a statement derived from perceived ideas but from experience and practice. The discussed ethos was and is evidently short-sighted as today's customers, supported in the field, are tomorrow's potential customers, ordering aircraft and airframe parts, this helping to shape future production programmes. The project group was cross-functional, and as team leader, it was my primary task to assure team members that quality improvement demanded a *no-blame concept* and that a complete *amnesty* was the way forward. This factor, in itself, was an acknowledgement of the widespread understanding that the manufacture of spares received low priority. Quality improvement is 'customer driven' and it could be said that customer support are at the cutting edge of customer care, in that we deal with the external customer who actually pays for products and services.
3 *Brainstorming*
   - Need for separate unit
   - Detail manufacture
   - Drawings – availability of same
   - Outside manufacture
   - Availability of confetti orders
   - Validation of sparable items and possible supply of next higher assembly
   - Jig and tool – possible medium-/long-term rectification though needs addressed immediately
   - Machined item/sub-assemblies
   - Attitude to spares/company culture
   - No formal system/policy.

   See Table 9.3.

Table 9.3 Prioritizing by criteria: Measure of level of difficulty

	Ease to solve *5 – easy* *1 – difficult*	*Ease of data collection* *5 – easy* *1 – difficult*	*Cost of poor support/supply* *5 – high* *1 – low*	*Comments*	*Total*
Need for separate unit	3,3,3,3	2,2,2,2	5,5,5,5	Refer to company security policy	29
Drawings supply	5,5,5,5	5,5,5,5	4,4,4,4	Register required in CSD.[a] Security policy	56
Outside manufacture	2,2,2,2	4,4,4,4	5,5,5,5	CSD to progress all sub-contract orders	44
Availability of orders	5,5,5,5	4,4,4,4	3,3,3,3	Order release pending stock check	37
Validation of order and supply of higher assembly	5,5,5,5	5,5,5,5	4,4,4,4	Ongoing	
Jigs and tools	1,1,1,1	2,2,2,2	5,5,5,5	Major group concern. Poor visibility	32

[a]CSD is Customer Service Department.

4 *Prioritizing by criteria*: Measure of level of difficulty.
5 *The remedial journey*: The group found 'brainstorming' to be the most effective and beneficial 'quality tool'. Ideas generated were as follows:
   - *Project status*: Spares manufacturing needs to be elevated to project status to ensure equality with production programmes.
   - *Cultural change*: Many mission statements are customer driven. We in customer support await the benefits of such a philosophy. The team believes that a cultural change towards 'spares' is developing within the company but also questions its origin. In other words, are spares getting a slightly higher profile due to uncertainty surrounding future production programmes? If demand on production increases, will customer/logistic support activities remain in the strategic plan?
   - *Awareness campaign*: Perhaps if there was a campaign to generate an awareness of both present and potential contribution spares and associated services make to the company/division business mix, greater priority would be given to same.
   - *Manufacturing strategy*: As the aircraft division is actively seeking future business, manufacturing's present strategy to off-load spares should be considered as business potential rather than sub-contracting.
   - *Present order book* – speculative ordering to meet perceived demand: CSO have already ensured that the present order book is realistic and speculative in a sense that is prudent to meet potential demand.
   - *The cost of poor quality*
     Loss of business thus loss of potential future customers.
     Reputation of the company – associated with poor support.
     An elementary skill of a successful business is maintaining present customer base. Poor support negates this.

6 *Cost of running the project and savings*: A significant factor which should be mentioned is that as a direct result of this project, CSO were able to reduce the monthly spares highlight list (a document which highlights firm customer orders outstanding in excess of 6 months), from a 6- to a 3-month threshold. This in itself is an indication that performance is improving. Three areas for consideration are:

   Savings due to the remedies
   Cost of running the project
   Cost of implementing remedies
   (a) *Savings due to the remedies*
      Tangibles
      - Order book reduced from 0.5 million to 340 000
      - Customer back orders reduced from 170 000 to 110 000.

Intangibles

- Time saved by closer functional co-operation which is effec tively 'process improvement' as output has improved an certain bottlenecks have been removed.
- Shortened lead times which have, in essence, brought i money more quickly.
- Cascade effect – the group believes that the project in itsel has created an awareness of the spares concept, althoug reservations must be placed regarding the origins o change, i.e. rundown of production programmes.

(b) *Cost of running the project*

Project team – 5 people
4 morning meetings 20 hours × 5 = 100 hours
8 1 hour meetings – 8 hours × 5 = 40 hours
Research, write-up and secretarial = 60 hours
200 hours × £8 per hour = £160

(c) *Cost of implementation*

The cost of implementation is negligible. No actual spend o introduction of new systems or methods. The project was vehicle to develop close coperation and awareness.

Conclusions

A quality culture requires everyone to focus on the customer, th ultimate user of output. Individuals need to understand who the cus tomer is and the work process performed to create the output Education and training is an integral part of the process of creatin the 'quality culture'. It is an ongoing activity requiring constant mon itoring and reinforcement.

Potential success

- Plan education and training, set targets
- Look down through the organization structure
- Select credible people as trainers
- Involve managers in training of their staff
- Involve senior managers to demonstrate commitment
- Encourage planning of application in the workplace

Potential failure

- Unrealistic timetables
- Inadequate delivery
- Desire for instant return on training investment
- Not training all employees
- Not allowing employees to complet training
- Inconsistent communication

Activity

Define the levels at which training and quality responsibilities are exercised in your organization. Then produce an analysis either:

- Demonstrating that training is a high-level and fully effective function in the organization *or*
- Concluding with recommendations ensuring that, in future, quality training responsibilities will be carried out at top management, line management, specialist and individual levels; and that training will assume its most appropriate role in the organization.

Summary points

- It is essential to base quality training plans on careful identification and analysis of training needs.
- The training process for employees at all levels should be regarded as one of empowerment.
- It is essential in the change process for everyone to have relevant knowledge or ability to use basic problem-solving tools.
- The effective use of quality-improvement tools is an integral aspect of continuous improvement activity.

References

1 Hamblin, A. C. (1974), *Evaluation and Control of Training*, Maidenhead: McGraw-Hill.
2 Lindahl, L. (1977), *Position and Change*, Boston, MA: Rendel Publishing.
3 Moon, C. G. and Hanton, T. (1968), 'Evaluating an appraisal and feedback training program', *Personnel*, November–December, 40.
4 Ishikawa, K. (1976), *Guide to Quality Control*, Tokyo: Asia Productivity Organization.

10 Assessment of quality

Introduction

To transform quality into a strategic business planning and manage-ment dimension, a new role must be created for the quality function. The quality function should be the organization's focal point and should gauge customer's quality expectations and subsequent satisfac-tion. Normally, organization systems are viewed in terms of interna dynamics as they apply to design, production, marketing, etc. It is now becoming essential that the interests of customers and suppliers are incorporated into the overall interests of an organization. If the quality function is to become a strategic business management function that helps to change a company's culture, the application of quality must be extended to all business processes and functions. On completion of this chapter you will:

- Recognize the importance of appraisal in the quality-improvement process
- Understand how to conduct an appraisal.

Strategic integration of quality

Throughout this text emphasis has been placed on the integration of quality into the organization's activities. Managing quality effectively is achieved through:

- Policy and strategy being formulated on the basis of information that is relevant to total quality (i.e. feedback from customers, suppliers, employees, etc.)
- The facilitation of quality improvement throughout the organization
- The development and maintenance of financial and non-financial quality measurement systems
- The promotion of 'partnership in quality' relations with customers and suppliers
- The planning and providing of quality skills education and training

In recent years many organizations have launched some form of quality initiative. As a result, some have become world-class competi-tors but others have not met with the same measure of success. Given that total quality is a cultural concept as well as being a highly systematic and systems-driven process, many organizations fail to understand quality as a strategic imperative which permeates all aspects of organizational performance.

As discussed, total quality is a dynamic set of ideas for action to manage and initiate change which is dependent on the determination of a continuous process of planning, doing and reviewing. The determination of a mission, visions and goals based on customer and supplier information leads to the identification of factors critical to the success of the organization in meeting customer requirements. Measurable objectives can then be established through the formulation of action plans which highlight responsibilities and allocation of resources. The planning phase should also address how the programme should be managed, how to involve employees; how to establish effective communications and the development of training plans (see Figure 10.1).

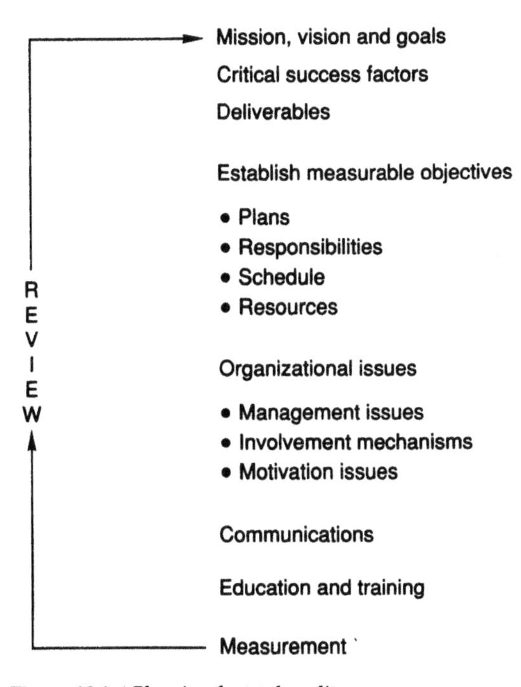

Figure 10.1 *Planning for total quality*

As the introduction and implementation of a total quality programme is a potentially complex operation, a quality strategy and associated plans must deliver measurable improvements if momentum is to be maintained. Furthermore, as total quality is a process of continuous improvement which involves change, quality ideas and actions must be owned by staff throughout the organization if the initiative is to have strategic impact.

Appraisal of quality

Figure 10.1 shows that measurement underpins all aspects of th improvement process. Throughout this book, measurement has bee an integral feature of quality activity. Quality audits, design review determining process capability, controlling variability, measuring qual ity costs, evaluating the effectiveness of teams and of training all con tribute to the:

- Establishment of current level of performance
- Highlighting of areas for improvement
- Monitoring of progress and achievement.

Quality appraisal is the process of identifying business practice attitudes and activities that are either enhancing or inhibiting th achievement of quality improvement within your organizatio Ideally, these factors would be recognized and addressed before quality improvement initiative is implemented. However, qualit appraisal adds great value at many points during the quality improve ment process:

1 At the start of an initiative
2 To guide action; during implementation
3 To pinpoint necessary adjustments
4 Or at any time thereafter, to benchmark progress. This can be don by external appraisal or by internal appraisal (i.e. self-assessment)

What are the benefits of quality appraisal?

During the quality-improvement process, particular activities rise t the top of the priority list only if the organization's senior managemen perceives benefit from them. However, the advantages of conducting quality appraisal are immediate, far-reaching and tangible.

List what you think are the main benefits of carrying out a qualit appraisal. Your answers may have included:

- *Supplying proof of need*: A properly structured quality appraisal pro vides factual evidence of the organization's true state of affair regarding quality and pinpoints those areas that would benefi most from the programme
- *Providing a baseline for future measurement*: When properly conducte the quality appraisal process produces a benchmark against which progress can be measured while implementing a quality-improve ment initiative. This enables an organization to monitor results and to identify any necessary mid-course corrections

- *Guiding management action*: By far the most powerful benefit of a quality appraisal is its ability to motivate senior management to identify and to commit themselves to actions required of them. Typically, much of the appraisal data point to systematic issues impeding quality improvement. These issues can be changed only through senior management's actions.

How to conduct a quality appraisal

The phases of the appraisal process include data gathering, data assimilation, management feedback and management action planning. The data-gathering phase requires particular attention since it sets the stage for the successful completion of the other phases. Any existing quality-specific organizational data should be utilized, and in cases where these do not exist or are incomplete, they must be collected. Interviews and surveys are cost-effective techniques for obtaining an accurate snapshot of the organization's status on important quality-improvement issues. These appraisal instruments should be designed so that data are collected in such a way so that the findings are actionable.

Understanding what questions to ask is essential to a successful quality appraisal. Many organizations have failed in this area because they have confused quality assessment with a more general employee-attitude survey. Although employee-attitude surveys can be useful and important, they do not serve the same purpose as a quality appraisal. An appraisal has only one objective: to identify business practices, attitudes, and activities that are barriers to quality improvement so that appropriate corrective actions can be taken (see Figure 10.2).

European framework for quality

The European Quality Award provides a framework against which progress and achievement can be mapped. It is not a standard but rather a model for self-assessment which provides a baseline for future measurement. Properly conducted, the quality assessment process produces a benchmark against which an organization's progress can be measured while implementing a quality-improvement initiative. It enables the organization to clearly identify its strengths and the areas in which approvements can be made using each criterion of the model.

The self-assessment process is more powerful than the traditional quality audit as it encompasses the entire activities of the organization rather than the quality system. The adoption of a strategy that includes self-assessment offers significant advantages to companies who wish to develop beyond the systems approach.

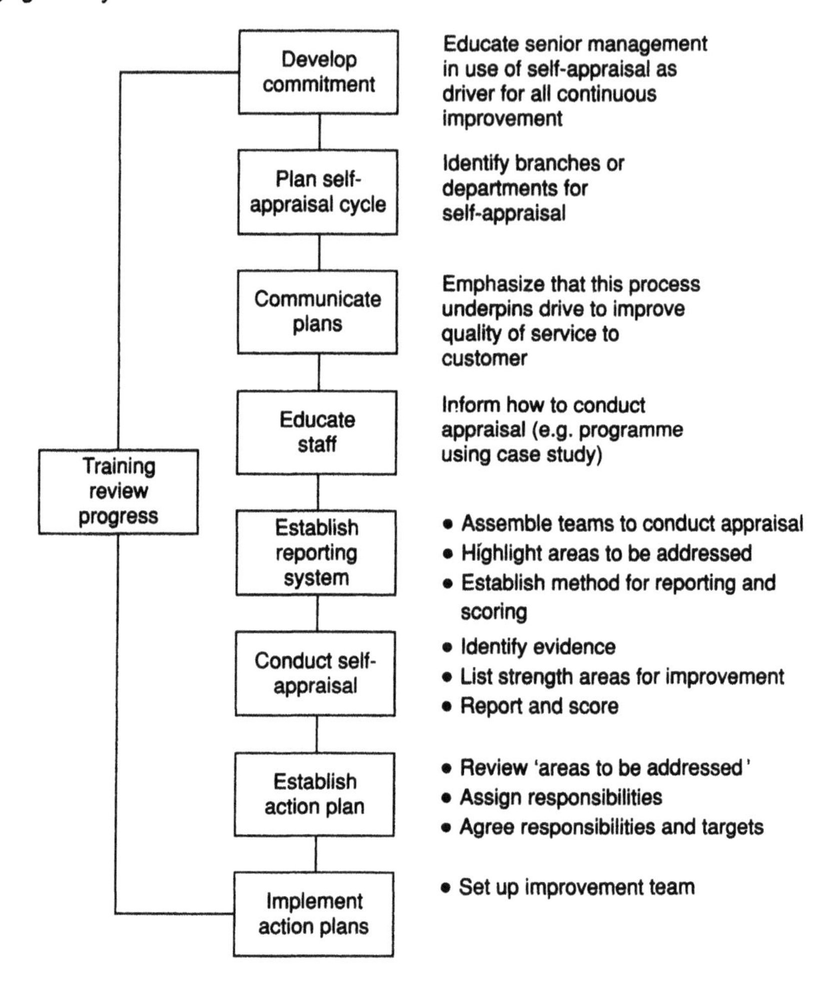

Figure 10.2 *Process of self-appraisal*

A simple appraisal technique

Many organizations, particularly if they are pursuing some kind of quality award, employ a third party to assist in appraisal. For example, to be accredited to the ISO quality management system it is necessary to be audited by a recognized assessor body. The appraisal activity described below is designed to act as a quick assessment tool covering in brief those areas which would be explored by a full appraisal. It can be undertaken at branch, section, department, division or organizational level.

There are a number of steps to be followed in carrying out the activity. You should read all these steps carefully before performing the appraisal to follow. These are:

1 Read the guidelines below for scoring. The guidelines are in two parts, one covering enablers criteria, the other, results criteria. Enablers criteria should be used to score against Units A-E below:
(a) Leadership
(b) Policy and strategy
(c) People management
(d) Resources
(e) Processes.
These address how results are being achieved. The results criteria should be used to score against Units F-I below:
(f) Customer satisfaction
(g) People satisfaction
(h) Impact on society
(i) Business results
These address what has been achieved.
2 Using these guidelines, attach a score to each of the elements within the nine units attached. A score of five would only be scored if *all* the areas highlighted under any one criteria were addressed. In some instances no score may be achieved if the area is not being addressed with branches, sections or organizations.
3 Having scored each criterion add up the total score for each element, then calculate the average score and list at each page.
4 The average score should then be plotted on the 'Radar Chart'. (Figure 10.3). The areas where you perform best will fall towards the outside of the chart.

Using the above steps complete the self-appraisal activity for your function, department or organization. Plot your answers on the Radar Chart. To do this properly you may need to carry out surveys, but for the purpose of this activity use information that is readily available. You may also need to consult other staff who have in-depth knowledge on certain subjects.

Review the weakest areas and prepare an action plan of two or three improvement points.

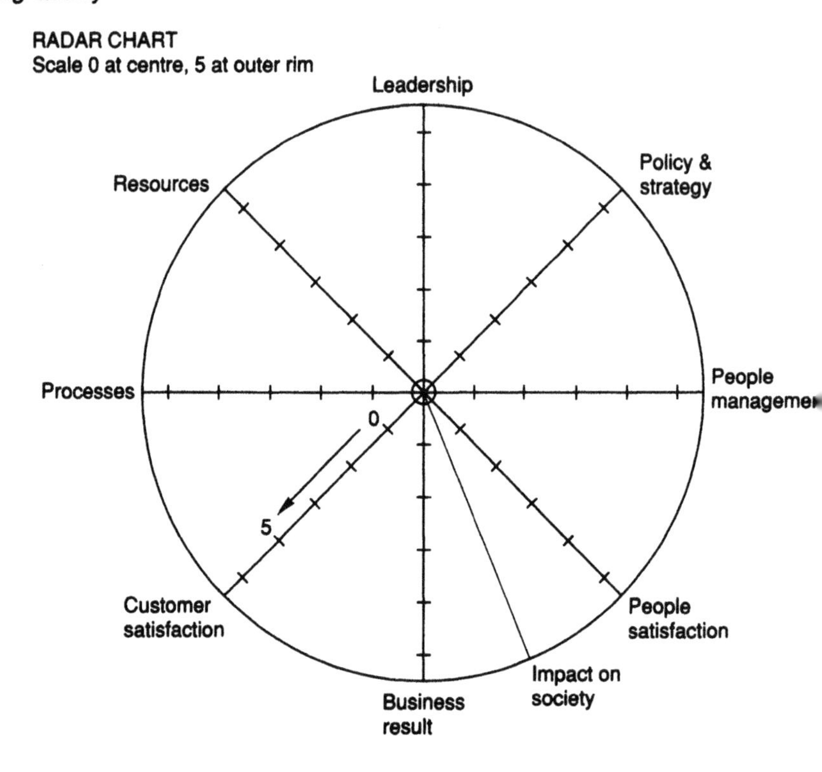

Areas for improvement

1 ____________________
2 ____________________
3 ____________________

Action points

1 ____________________
2 ____________________
3 ____________________

Figure 10.3 *Total quality health check*

GUIDELINES
Enablers scoring criteria

0	Not addressed within branch, department, organization
1	Little effective use
2	Applied to about one-quarter of the potential when considering all relevant areas and activities. Activities are subject to occasional review
3	Applied to about half of the potential when considering all relevant areas and activities. Activities are subject to regular review. Some evidence of systematic approach
4	Applied to about three-quarters of the potential when considering all relevant areas and activities. Evidence that continuous review contributes to branch, departmental or organizational effectiveness. Approach integrated into normal operations
5	Applied to full potential in all relevant areas and activities. Approach totally integrated into normal working patterns.

Results scoring criteria

0	Not addressed within branch, department, organization
1	Results address few relevant areas and activities
2	Results show some positive trends and comparison with own targets. Address some relevant areas and activities
3	Results show positive trends over the last three years. Favourable comparisons with own targets. Address many relevant areas and activities
4	Most results show positive trends over at least three years. Favourable comparisons with own targets and comparison with external organizations. Results address most relevant areas and activities
5	Very positive trends in all areas over at least three years. Excellent comparisons with own targets. Results address all relevant areas

Unit A Leadership: The behaviour of all managers in driving the organization towards total quality

	Score
Demonstrated management commitment to total quality	
Managers lead by example and act as role models	
Participation in the review and progress of total quality in the organization	
Manager's involvement in providing timely recognition to individuals and teams	
Provision of total quality training and management participation in training	
Provision of time and resources in support of continuous improvement	
Involvement with 'customers' in improvement activity	
Involvement with 'suppliers' in improvement activity	
Total score	
Average = total/8 =	

Unit B Policy and strategy: The organization's mission, values, vision and strategic direction and ways in which it achieves them

	Score
Is the concept of total quality reflected in the organization's values, vision and mission?	
Does the organization's strategy embrace the concepts of total quality?	
Is the policy and strategy based on feedback from the organization's people?	
Is the policy and strategy based on data from competitors and 'best in class'?	
How well is the business plan aligned with the policies?	
How effectively have policy and strategy been communicated?	
Total score	
Average = total/6 =	

Unit C People management: How the organization releases the full potential of its people to continuously improve the business

	Score
The use of surveys to obtain the perceptions of employees	
Establishment and implementation of training plans	
Review and improvement in effectiveness of training received	
The use and effectiveness of the suggestion scheme	
The use and effectiveness of quality improvement teams	
The use of two-way briefing meetings	
Deployment of organizational information	
Total score	
Average = total/7 =	

Unit D Resources: How the organization's resources are effectively deployed in support of policy and strategy

	Score
Effective use of 'quality cost' information for continuous improvement	
Management information validity and accuracy	
Continuous improvement of information systems and data presentation	
Effective utilization of new technology	
Development of people's skills and capabilitites harmonized with technology development	
Total score	
Average = total/5 =	

Unit E Processes: How processes are identified, reviewed and, if necessary, revised to ensure continuous improvement of the organization's business

	Score
Definition of critical processes and methodology for improvement of processes	
Establishment of process ownership and agreed standards of operations	
Use of performance measures in managing processes	
Level of innovation and creativity in process improvement	
Effective communication of process improvements	
Auditing of process changes to ensure that results are achieved	
Total score	
Average = total/6 =	

Unit F Customer satisfaction: What the perception of external customers is of the organization and its products and services

	Score
Proactive in dealing with customer problems	
Simplicity, convenience and accuracy of customer documentation	
Awareness of customer problems and complaint handling	
Capability of meeting customer's specifications	
Standard and responsiveness of technical support	
Responsiveness and flexibility of organization to customer demands	
Total score	
Average = total/6 =	

Unit G People satisfaction: What the people's feelings are about their organization

	Score
Standard of working environment and amenities	
Standard of health and safety provisions	
Standard of employee communication at local and organizational level	
Provision of training, development and retraining for all employees	
Level of involvement in total quality process	
Awareness of values, vision and strategy	
Process of appraisal, target setting and career planning	
Level and trend in absenteeism, staff turnover and grievances	
Total score	
Average = total/8 =	

Unit H Impact on society: What the perception of the organization is among society at large

	Score
Contribution to charity	
Support for education and training	
Support for medical and welfare	
Support for sports and leisure	
Conservation of energy and raw materials	
Number of public complaints	
Contribution to the local economy	
Total score	
Average = total/7 =	

Unit I Business results: What the organization is achieving in relation to its planned business performances

	Score
Achievement and trends in meeting budget targets	
Improvement in value-added per staff	
Reduction trends in waste, non-conformance, non-value-added activity (i.e. claim-processing time)	
Improvements in service level achievements targets	
Total score	
Average = total/4 =	

SELF-APPRAISAL CHECK

RADAR CHART

Scale 0 at centre, 5 at outer rim (see Figure 10.3)

AREAS FOR IMPROVEMENT

1 ______________________________

____

2 ______________________________

____

3 ______________________________

____

ACTION POINTS

1 ______________________________

____

2 ______________________________

____

3 ______________________________

____

The following case study illustrates how a small manufacturing company has applied self-appraisal to monitor progress towards total quality. The areas for improvement identified have been incorporated into developing strategic plans which include total quality objectives.

Electron Engineering

Rationale for quality programme

Although the company remained both busy and profitable through the 1980s the managing director recognized that the business environment was changing: customer expectations and competition were increasing. In December 1989, realizing that continued success would require a new, more proactive way of managing the business, he attended a total quality workshop. Although the concept was new, he recognized its potential and arranged a complete quality audit of the company in an effort to determine where to focus effort.

The quality audit was completed in April 1990 and indicated that the company had a positive image in the marketplace and that it was, on the whole, quality conscious. However, the management structure was insufficiently developed and no formal and systematic quality system was in existence. Furthermore, although the company was profitable, it was recognized that this could be significantly improved by better planning and the introduction of a management approach that increased employee involvement and motivation.

The diversity of problems revealed by the audit convinced manage ment that an approach that embraced all aspects of the company' activities would be required. Therefore in 1990 the company embarke on a total quality management programme.

Getting started

The company began by forming a project management team (PMT) t take responsibility for implementing and maintaining the programm They began by establishing 34 quality objectives, embodied by six mai policy statements. These statements reflected the need to develop quality culture within the company as an important part of everyda business activity, as well as part of strategic planning activity. Th policies focused on business planning, performance monitoring an improving employee, customer and supplier relationships.

Management recognized that their leadership of the programme ha to be highly visible. They began by actively encouraging all employee to identify opportunities for improvement and to become involved i team activities. They also participated in PMT and action team meet ings. Furthermore, they began to hold monthly team meetings to dis cuss business strategy, performance against targets and problems. Thi communication process proved to be both efficient and effective an has enabled employees to raise concerns and discuss other items o interest.

Implementation

The implementation phase began in August 1990 with a brainstormin session involving all employees. This focused on highlighting problen areas and was facilitated by a consultant employed through th Advisory Service to Industry grant assistance scheme. The ideas gen erated were grouped by the PMT and three action teams were forme to address them. These consisted of employees from all levels an functional areas, and a member of the management team.

Electron Engineering firmly believe that teamwork is an essentia component of the implementation of TQM. A team approach ha been used to build trust, improve communications and to hel develop an interdependence between departments and individuals. A number of action teams have been formed over the past two years t address specific tasks, and this approach has helped to create a wor environment where ability and initiative are encouraged, recognize and rewarded.

Quality system standards

Electron recognized the importance of process management at the outse and during the early implementation phase of the programme began t flow chart processes. This enabled them to identify opportunities for improvement and ways to improve process management.

The need to manage processes more effectively combined with the recognition that the market would eventually demand evidence of quality capability prompted the company to seek ISO 9002 certification. It is interesting that this was just one of the 34 objectives identified during the planning phase of the programme and was not seen as the main objective or as an end in itself. Furthermore, it emphasizes that the quality system is only a part of their TQM approach.

Training and development

The company realized that training and education were vital if the whole-hearted participation of all employees in the continuous improvement programme was to be secured. This is reflected in the fact that nine of the objectives derived from the organization's policy statements related to training and employee development.

In November 1990 a training needs analysis was conducted which recommended training requirements for each employee, including management. The PMT took responsibility for ensuring that these needs were met and training began with all employees attending a quality-awareness/problem-solving workshop. However, it was quickly realized that further training could be facilitated by the management team. As a result, the majority of training has taken place during action team meetings. Additional training needs have been identified during these meetings and regular seminars are held to train staff in new developments (e.g. the latest design and manufacturing techniques).

Individual development needs are reviewed on an annual basis as part of an employee-appraisal process. Quality has been built into this process through the drafting of individual job descriptions to include quality-related obligations. Although these focus mainly on the quality system, the company intend to extend them to total quality-related obligations. The appraisal process ensures that employees know what their objectives are and is another communications mechanism.

Customer focus

Customer focus and satisfaction to Electron Engineering leads to repeat business and referrals, and derives from all contact the customer has with the business. It has therefore become the company's number-one agenda item.

Telephone sales and enquiry procedures have been developed to capture customer information. An enquiry follow-up is begun a week after each quotation to determine if Electron have won or lost the order, the reasons why, and to help build a picture of the organization's competitive position. This information has enabled the company t forecast the percentage of orders lost, pending or secured, and ha provided invaluable information for sales budget preparation and pro duction planning.

An enquiry follow-up report of the entire customer database is pro duced twice weekly to keep the management team informed abou how they can continue to understand and meet customer require ments. Potential or lost customers can be identified at a 'glance', free ing resources to concentrate on winning them by tuning the 'tota product' to their needs.

This approach has enabled Electron to be more proactive in thei sales and marketing efforts, and to be flexible enough to cope wit changing market requirements. Good relationships are maintaine with both customers and suppliers, and feedback is both welcome and respected since it is a reflection of the company's success in attain ing its quality goals.

Company self-assessment process

In order to quantify fully the precise status of the organization i respect of continuous improvement activities, an assessment o Electron Engineering was conducted, addressing the organization a a whole. The mechanism employed was the European Quality Awar Model for Self-Appraisal, the primary objective of which is the regula and systematic review of an organization's activities and results.[1]

This process allows the organization to clearly discern its strength and areas in which improvements can be made. There are nine criteri used to appraise an organization's progress towards total quality man agement:

- Leadership
- Policy and strategy
- People management
- Resources
- Processes
- Customer satisfaction
- People satisfaction
- Impact on society
- Business results

Leadership

This criterion assesses the behaviour of all managers in driving a organization towards total quality and evidence is required to demon strate involvement in leading quality management. The findings were

- No evidence of a formal total quality implementation programme

- Unstructured approach to training
- No process of benchmarking
- Lack of evidence of how management have funded learning
- No formal evidence of customer requirements

Policy and strategy

This criterion addresses the company's values, vision and strategic direction and the manner in which it achieves them. Evidence is needed of how policy and strategy are based on the concepts of total quality. The findings were:

- No strategy for attainment of each objective
- No medium-/long-term strategy or annual plan
- Need to integrate total quality activities with needs of the business
- No formal method of evaluating effectiveness/relevance of policy in terms of customer requirements, competitor analysis, benchmarking, etc.

People management

This criterion addresses the management of the company's people in respect of how the full potential of individuals is released to improve its business continuously. Evidence is required of: development of core skills through recruitment, training and career progression; how performance targets are agreed and reviewed; the promotion of involvement and the extent of improvement. The findings were:

- Lack of formal mechanism for determining employee perceptions of the organization on an ongoing basis
- No evidence of an up-to-date training needs analysis based on business requirements
- Team activities not reviewed for effectiveness
- Annual appraisal does not review total quality objectives, activities.

Resources

This criterion addressed the management, utilization and presentation of resources. Evidence is needed of how the company improves its business continuously by optimization of financial, information and technological resources. The findings were:

- No link between total quality programme and financial/budgetary provision
- Financial benefits of total quality not measured in full
- Financial reporting system not linked to other systems

- No process for evaluating and monitoring technology application t identify areas for improvement.

Processes

This criterion assesses the management of all the value-adding activ ities within an organization. Evidence is required of how key processe are identified; how the company systematically manages its key anc support processes; how process performance parameters are used tc review key processes; and how process changes are implemented anc evaluated. The findings were:

- Lack of evidence of systematic identification, prioritization anc review of key processes
- No measures, standards or targets exist to facilitate process manage ment
- Process improvements have not been qualified.

Customer satisfaction

This criterion assesses what the perception of external customers is o an organization, its products and services. Evidence is required of the company's success in satisfying the needs and expectations of custo mers. The findings were:

- Absence of a formal process for measuring perception of externa customers
- Absence of trend data, comparison with targets, competitors, etc.

People satisfaction

This criterion assesses the attitudes and feelings of the employees of ar organization. Evidence is needed of the company's success in satisfying the needs and expectations of its people. The findings were:

- No measures and trends of people satisfaction exist, comparison with targets, competitors and best-in-class.

Impact on society

This criterion assesses what the perception of an organization is among the community at large, in respect of the company's approach to quality of life, the environment and to the preservation of global resources. The findings were:

- Lack of an 'impact on society' policy
- No specific measures or trend data exist.

Business results

Since embarking upon a total quality programme, the effect on the organization has been enormous, and impressive results have been achieved in many areas:

- Percentage of enquiries priced and confirmed to customers on time has increased from 25% to 98%
- Increased shopfloor capacity, due to the elimination of wasteful processes
- Value output per employee has doubled
- More accurate accounts of 'man-hours' available at any given time
- Accurate and constantly updated stock valuations, facilitating stock reductions of over 30%
- An increased profitability ration.

Conclusions of case study

The policies and objectives of Electron Engineering which focus on quality, total commitment to continuous improvement and a vision for the future success of the organization are gradually yielding return. A concentrated effort to focus on continuous improvement on all levels rather than on procedural compliance has been accepted as the norm within the organization.

The self-appraisal process is an indication of commitment. As the application criteria for the European Quality Award stipulate that evidence be provided of the management of the nine criteria which demonstrates trends and measurement of performance, the scope of the award is considered to be more relevant to larger organizations. However, Electron Engineering recognize that the process of self-appraisal on an organization-wide scale is of considerable benefit in aiding the monitoring of actual progress and in determining policies and associated action plans based on the areas for improvement identified.

Conclusions

The use of an appraisal framework such as the European Quality Award avoids the trap of focusing exclusively on quality-related outputs and results. Instead, it forms a thorough introspective examination of the business processes and management support systems through which quality is achieved. Quality assessment is the most effective way to identify systematic quality problems. For example, in the midst of an improvement project one organization discovered that continued quality training would be futile unless the cross-functional business processes, identified through a quality appraisal, were addressed in a systematic way. Following an analysis of their business practices management discovered that barriers existed to applying the quality improvement skills that staff were being taught. Once the management team accepted responsibility for these barriers, improvement resulted.

Quality appraisal is also an effective way of involving senior management in the improvement process. Since appraisal, by its nature uncovers systemic barriers to improvement, senior management can become actively involved in leading the change process.

Summary points

The effective management of quality requires:

- Customer-first orientation.
- Top management leadership of the quality-improvement process.
- Active involvement of employees.
- Reduction of product and process variation.
- Provision of ongoing education and training of employees.
- Emphasis on prevention rather than detection.
- Performance measures that are consistent with the goals of the organization.
- Emphasis of product and service quality in design.
- Cooperation and involvement of all functions within an organization.
- Awareness of the needs of internal customers.

Reference

1 European Foundation for Quality Management (1993), *The European Model for Self-Appraisal.*

Bibliography

Albrecht, K. and Bradford, L. F. (1990), *The Service Advantage: How to identify and fulfil customer needs*, Homewood, Illinois: Dow Jones-Irwin.

Anundsen, K. (1979), 'Building teamwork and avoiding backlash', *Managerial Review*.

Atkinson, P. E. (1990), *Creating Culture Change: The key to successful total quality management*, Bedford: IFS Publications.

Aubrey, C. A. (1988), *Quality Management in Financial Services*, Wheaton: Hitchcock.

Baker, K. R. (1974), *Introduction to Sequencing and Scheduling*, New York: Wiley.

Belbin, R. M. (1981), *Management Teams, Why They Succeed or Fail*, Oxford: Butterworth-Heinemann.

Berry, L. L., Bennett, D. R. and Brown, C. W. (1989), *Service Quality: A Profit Strategy for Financial Institutions*, Homewood, Illinois: Dow Jones-Irwin.

Blumberg, D. F. (1991), *Managing Service as a Strategic Profit Centre*, New York: McGraw-Hill.

Bone, D. and Griggs, R. (1989), *Quality at Work*, London: Kogan Page.

Bower, M. (1972), *Managing the Resource Allocation Process*: Harvard Business.

Brown, A. (1989), *Customer Care Management*, Oxford: Butterworth-Heinemann.

Buzzell, R. and Gale, B. T. (1987), *The PIMS Principles – Linking Strategy to Performance*, New York: The Free Press.

BS 5750: Part 1: 1987, Specification for Design/Development, Production, Installation and Servicing.

BS 5750: Part 2: 1987, Specification for Production and Installation.

BS 5750: Part 3: 1987, Specification for Final Inspection and Test.

BS 5750: Part 4: 1990, Guide to the use of BS 5750: Part 1, Specification for Design, Manufacturing and Installation. Guide to the use of BS 5750: Part 2, Specification for Manufacture and Installation. Guide to the use of BS 5750: Part 3, Specification for Final Inspection and Test.

BS 6143: Part 1: 1992, Guide to the Economics of Quality: Part 1, Process Cost Model.

BS 6143: Part 2: 1990, Guide to the Economics of Quality: Part 2, Prevention, Appraisal and Failure Model.

BS 7850: Part 1: 1992, Guide to Management Principles, Part 2: 1992, Guide to Quality Improvement Methods, Total Quality Management.

Caplan, F. (1990), *The Quality System* (2nd edition) Pennsylvania, PA Chilton Book Company.
Cooper, R. and Kaplan, R. S. (1991), *The Design of Cost Managemen System*, Englewood Cliffs, NJ: Prentice-Hall.
Crosby, P. B. (1979), *Quality is Free*, New York: McGraw-Hill.
Crosby, P. B. (1986), *Running Things, The Art of Making Things Happen* Singapore: McGraw-Hill.
Crosby, P. B. (1987), *Quality Without Tears* (3rd edition) Singapore McGraw-Hill.
D'Egidio, F. (1990), *The Service Era Leadership in a Global Environment* Cambridge, Massachusetts: Productivity Press.
Davidow, W. H. and Uttal, B. (1989), *Total Customer Service the Ultimat Weapon*, New York: Harper & Row.
Deming, W. E. (1980), *Out of the Crisis*, Cambridge, Massachusetts MIT, Centre for Advanced Engineering Study.
Deming, W. E. (1986), *Out of the Crisis*, Cambridge, Massachusetts Institute of Technology.
Deming, W. E. (1982), *Quality, Productivity and Competitive Position* Cambridge, Massachusetts: Institute of Technology.
Desatnick, R. L. (1988), *Managing to Keep the Customer*, San Francisco CA: Jossey-Bass.
DiPrimio, A. (1987), *Quality Assurance in Service Organisations* Pennsylvania: Chiltern Book Company.
European Foundation for Quality Management (1993), *The Europea Model for Self-Appraisal*.
European Organization for Quality (1991), *Proceedings of 35th Annua Conference. 'The Human Factor in Quality Management'*, Prague: EOQ
Feigenbaum, A. V. (1983), *Total Quality Control* (3rd edition), New York McGraw-Hill.
Gitlow, H. S. and Gitlow, S. J. (1987), *The Deming Guide to Quality an Competitive Position*, Englewood Cliffs, NJ: Prentice-Hall.
Griffin, R. W. (1990), *Management*, (3rd edition), New York: Houghton Mifflin.
Hamblin, A. C. (1974), *Evaluation and Control of Training*, London McGraw-Hill.
Harrington, H. J. (1987), *The Improvement Process, How America's Leading Companies Improve Quality*, New York: McGraw-Hill.
Hellriegel, D., Slocum, J., Woodman R. (1989), *Organisational Behaviour* St Paul, Minnesota: West.
Heskett, J. L., Sasser, W. E. and Hart, C. W. (1990), *Servic Breakthroughs, Changing the Rules of the Game*, New York: The Fre Press.
Hickman, C. R. and Silva, M. A. (1986), *Creating Excellence*, London Unwin.
Hillman, P. (1992), 'It's simple, but not easy', *Managing Service Quality* September.

Ishikawa, K. (1976), *Guide to Quality Control*, Tokyo: Asia Productivity Organization.
ISO (1991), *Quality Management and Quality System Elements Part 2. Guidelines for Services (ISO 9004-2)*, Geneva.
Juran, J. M., Gryna, F. M. and Bingham, R. S. (eds) (1979), *Quality Control Handbook*, (3rd edition), New York: McGraw-Hill.
Juran, J. M. and Gryna, F. M. (1980), *Quality Planning and Analysis*, New York: McGraw-Hill.
Katz, B. (1987), *How to Turn Customer Service into Customer Sales*, Aldershot: Gower.
Kroeger, O. and Thuesen, J. M. (1992), *Type Talk at Work*, Delacorte Press.
Kubler-Ross, E. (1986), 'On death and dying', *Practitioner*, **18**, December.
Lammermeyr, H. U. (1990), *Human Relations the Key to Quality*, Wisconsin: ASQC Quality Press.
Lapin, L. (1988), *Quantitative Methods for Business Decisions*, Orlando, Florida: Harcourt Brace Jovanovich.
Larson, C. E. and Lafasto, F. M. J. (1989), *Teamwork, What Must be Right/What Can Go Wrong*, Beverly Hills, CA: Sage.
Latsko, W. J. (1986), *Quality and Productivity for Bankers and Financial Managers*, New York: Marcel Dekker.
Lindahl, L. (1977), *Position and Change*, Boston, MA: Rendel Publishing.
Lipsey, P. (1963), *An Introduction to Positive Economics*, London: Weidenfeld and Nicholson.
Liswood, L. A. (1990), *Serving Them Right*, New York: Harper Business.
Lock, D. and Smith, D. J. (eds) (1990), *Gower Handbook of Quality Management*, Aldershot: Gower.
Love, J. F. (1988), *McDonald's: Behind the Arches*, London: Batman.
Margulies, N. (1973), *Organisational Change, Techniques and Application*, Illinois: Scott-Foresman.
Martin, W. B. (1989), *Managing Quality Customer Service*, London: Kogan Page.
Murphy, J. A. (1988), *Quality in Practice*, Dublin: Gill and Macmillan.
Oakland, J. S. (1989), *Total Quality Management*, Oxford: Butterworth-Heinemann.
Peel, M. (1987), *Customer Service, How to Achieve Total Customer Satisfaction*, London: Kogan Page.
Peters, T. (1986), *Quality!*, Palo Alto, CA: Peters Organisation.
Peters, T. (1987), *Thriving on Chaos*, London: Macmillan.
Peters, T. J. and Waterman, R. H. (1982), *In Search of Excellence: Lessons from America's Best Run Companies*, New York: Harper & Row.
Price, F. (1984), *Right First Time*, Aldershot: Gower.
Reichheld, F. F. and Sasser Jr, W. E. (Sept./Oct. 1990, Jan./Feb. 1991), 'Zero defections: Quality Comes to Services', *Harvard Business Review*.

Robbins, S. P. (1992), *Essentials of Organisational Behaviour*, (3rd edition), Englewood Cliffs, NJ: Prentice-Hall.

Rosander, A. C. (1985), *Applications of Quality Control in the Service Industries*, Wisconsin: ASQC Quality Press.

Sayle, A. J. (1988), *Management Audits: The Assessment of Quality Management Systems* (2nd edition), London: Allan J Sayle.

Schein, E. H. (1989), *Organisational Culture and Leadership*, San Francisco, CA: Josey-Bass.

Schonberger, R. J. (1990), *Building a Chain of Customers*, New York: The Free Press.

Starr, M. K. (1989), *Managing Production and Operations*, Englewood Cliffs, NJ: Prentice Hall.

Thomas, K. (1976), *Conflict and Conflict Management*, Handbook of Industrial and Organisational Psychology, Chichester: Wiley.

Townsend, P. L. and Gebhardt, J. E. (1986), *Commit to Quality*, New York: Wiley.

Wilkie, W. L. (1986), *Consumer Behaviour*, New York: Wiley.

Woodcock, M. (1989), *Team Development Manual*, (2nd edition), Aldershot: Gower.

Wright, P. and Taylor, D. (1984), *Improving Leadership Performance – A Practical Approach to Leadership*, Englewood Cliffs, NJ: Prentice-Hall.

Index

9 780750 618236

For Product Safety Concerns and Information please contact our EU representative GPSR@taylorandfrancis.com Taylor & Francis Verlag GmbH, Kaufingerstraße 24, 80331 München, Germany

Batch number: 08158361

Printed by Printforce, the Netherlands